The Politics of Scale

The Politics of Scale

A History of Rangeland Science

NATHAN F. SAYRE

The University of Chicago Press
Chicago and London

The University of Chicago Press, Chicago 60637
The University of Chicago Press, Ltd., London

Published 2017.
Printed in the United States of America

26 25 24 23 22 21 20 19 18 17   1 2 3 4 5

ISBN-13: 978-0-226-08311-7 (cloth)
ISBN-13: 978-0-226-08325-4 (paper)
ISBN-13: 978-0-226-08339-1 (e-book)
DOI: 10.7208/chicago/9780226083391.001.0001

Library of Congress Cataloging-in-Publication Data
Names: Sayre, Nathan Freeman, author.
Title: The politics of scale : a history of rangeland science / Nathan F. Sayre.
Description: Chicago : The University of Chicago Press, 2017. |
Includes bibliographical references and index.
Identifiers: LCCN 2016033176 | ISBN 9780226083117 (cloth : alk. paper) |
ISBN 9780226083254 (pbk. : alk. paper) | ISBN 9780226083391 (e-book)
Subjects: LCSH: Rangelands—United States—History. | Range ecology—
United States. | Range ecology—Economic aspects—United States. |
Range policy—United States—History.
Classification: LCC SF85.3 .S29 2017 | DDC 636.08/45—dc23 LC record
available at http://lccn.loc.gov/2016033176

♾ This paper meets the requirements of ANSI/NISO Z39.48-1992
(Permanence of Paper).

To Henry, Lila, Anastasia, and Robin,
and to the memory of
Robert Freeman Sayre (1933–2014)

CONTENTS

A NOTE ON UNITS OF MEASUREMENT

Imperial, US customary, and metric units of measurement are all found in the text that follows, reflecting the practices of various people and institutions at different places and times. Given the text's heavy reliance on direct quotations from primary materials, I have chosen not to convert measurements into a single system, nor to insert conversions in parentheses throughout the narrative. Readers may wish to refer back to these conversion factors as they proceed:

1 acre = 0.4 hectare
1 hectare = 2.47 acres
1 square mile = 640 acres = 1 "section" = 2.59 square kilometers
1 square kilometer = 247 acres = 0.39 square miles
1 pound = 0.45 kilograms
1 kilogram = 2.2 pounds
1 pound per acre = 1.12 kilograms per hectare
1 kilogram per hectare = 0.89 pound per acre

Map 0.1. The western United States, showing locations of places mentioned in the text. The 100th meridian, also shown, is the rough western boundary of feasible crop agriculture without irrigation. By Darin Jensen and Alicia Cowart.

PREFACE

Rangelands are embattled, imperiled, and poorly understood in the United States and around the world. In places across Africa, Asia, Australia, Europe, and the Americas, governments and developers endeavor to install farms, mines, suburbs, dams, energy and infrastructure projects, parks, and preserves on rangelands heretofore used by pastoralists or ranchers. Many environmental groups decry both the developers and the livestock producers, fearing fragmentation, loss of habitat, and "desertification." In the western United States, disputes over federally owned rangelands have flared and smoldered for over a century with little progress; the demands of the militants who occupied the Malheur National Wildlife Refuge in Oregon in the weeks before this book went to press, for example, were nearly identical to those made by livestock associations in the 1940s. Scientific claims are routinely invoked to support the positions of one or more parties in these political and economic struggles. But the science of rangelands—what is known and unknown, where and how such knowledge was produced, and which claims it really supports—is a veritable black box for almost everyone. This book opens up that box.

Inside, we find a sort of looking glass that both reflects and disguises the conflicts that rangeland science is supposed to transcend and thereby help resolve. From the outset, the politics of land use and degradation and the ambition to profit from rangelands shaped not only how scientists worked but the very concepts with which they constructed their ideas. Then a flawed theory was imposed on scientists working for the United States Forest Service, the agency that dominated the young field. In turn, their research was interpreted as verifying the theory, notwithstanding their own misgivings and growing evidence to the contrary. The flaws in the theory disappeared through the looking glass, leaving partisans on all sides—environmentalists, ranchers, government agencies, and even range scientists—blind to the fact that they were trapped in a set of concepts and ideas that could only perpetuate their disputes. In recent decades, rangeland scientists have overthrown the old theory, but most rangeland conflicts are still fought in (and on) its terms. My hope is that illuminating this history may help break the vicious cycle and enable more constructive ways forward.

Many people have assisted and encouraged me in producing this book. It began—at least in retrospect—on a cold winter night in the early 2000s at the Saddle and Spur Tavern in Douglas, Arizona, when Curt Meine returned my questions with a blunt and friendly admonition: "No one has written a history of range science. You have to do it!" Over the ensuing years, Curt has been both a valuable source of information and feedback and an inspiring example of engaged, public scholarship.

The gathering that brought Curt and me together in Douglas was the annual science conference of the Malpai Borderlands Group, an organization of ranchers, scientists, conservationists, and government agency employees dedicated to preserving "working wilderness" in the rangelands of southeastern Arizona and southwestern New Mexico. In working with Malpai and its cooperators for more than a decade, I have accumulated countless debts of gratitude, hospitality, and wisdom. My thanks to Bill and Mary McDonald, Ben Brown, Peter Warren, Ray Turner, Warner Glenn, John Cook, Mike Dennis, Edward Elbrock, Rick and Heather Knight, Bill Radke, Ron Bemis, Don Decker, Reese Woodling, Carl Edminster, Gerry Gottfried, Larry Allen, Kelly Cash, Steve Spangle, Bill Lehman, Stephen Williams, Pete Sundt, Myles Traphagen, David Hodges, Gita Bodner, Sam Smith, Jennifer Medina, Donnie Kimble, Billy Darnell, Dave Dunagan, Bill Miller, Jay Dusard, and the Hadley, Krentz, Magoffin, Winkler, and Austin families. A special salute to three Malpai friends who sadly are no longer with us: Wendy Glenn, Drum Hadley, and Rob Krentz.

Outside of my family, no one has done more to help me complete this project than Kris Havstad, whose knowledge of rangeland ecology and US range administration is exceeded only by his magnanimity and sense of humor. As supervisory scientist of the USDA Agricultural Research Service's Jornada Experimental Range, Kris gave me the scientific papers I needed—a stack about a foot tall, back in the days before pdfs—to write *The New Ranch Handbook,* and then he gave me a postdoctoral fellowship to bring social scientific dimensions into the Jornada's research programs. The many extraordinary staff and scientists at the Jornada—including Brandon Bestelmeyer, Joel

Brown, Bernice Gamboa, Valerie LaPlante, Walt Whitford, Jeff Herrick, Deb Peters, Al Rango, Eldon Ayers, Dawn Browning, Rhonda Skaggs, Alfredo Gonzalez, Ed Fredrickson, Michaela Buenemann, Matt Levi, Mike Duniway, and Greg Okin—have been both generous colleagues and good friends. Kris subsequently invited me to participate in the Jornada Basin Long-Term Ecological Research program, which led to fruitful collaborations with Gary Kofinas, Laura Ogden, Gil Pontius, Hannah Gosnell, and the rest of the Maps and Locals project. Finally, he helped fund travel to the National Archives in College Park and Denver, and he generously shared the Jornada archives for this project.

Beyond Malpai and Jornada, I have enjoyed the company and benefited from the knowledge of scores of people who share a love of southwestern rangelands: George Ruyle, Tom Sheridan, Mitch McClaran, Lamar Smith, Steve Archer, Chuck Hutchinson, Guy MacPherson, Joe Wilder, David Yetman, Bob Steidl, Bill Shaw, and Phil Ogden with the University of Arizona; Courtney White and Genn Head, Kirk and Tamara Gadzia, Bill Zeedyk and the extended Quivira Coalition community in Santa Fe; Dan Robinett, now retired from the Natural Resources Conservation Service; Dave Stewart, Jennifer Ruyle, Frank Hayes, Dave Bradford, Paula Medlock, and Ed Holloway in the Forest Service; Mike Goddard, Marty Tuegel, and Sally Flatland in the US Fish and Wildlife Service; Julia Fonseca and Linda Mayro of Pima County; the King, Miller, Rowley, Chilton, and Kasulaitis families in the Altar Valley; Mac Donaldson, Dennis and Deb Moroney, Walt Myers, Barbara Clark, Harry Greene, Ross and Susan Humphreys, Linda Kennedy, Sid and Cheryl Goodloe, Ed Marston, David Ogilvie, Conrad Bahre, and David Remley.

In the course of research and writing, I have been fortunate to receive questions and suggestions from audiences at numerous conferences, colloquiums, and workshops where I presented work in progress. My thanks to the organizers and attendees at the Department of Geography at the University of Colorado Boulder, the Yale School of Forestry and Environmental Studies, the Environment and Society Department at Utah State University, the Energy and Resources Group at the University of California Berkeley, and at meetings of the Society for Range Management, the Ecological Society of America, the American Society for Environmental History, the Society for Conservation Biology, the California Native Grasslands Association, the Jornada Symposium, the Malpai science meetings, and the American Association of Geographers. Special thanks to Rod Neumann and the students in his graduate seminar at Florida International University, who read the entire manuscript in fall 2015 and provided a wealth of insightful and useful suggestions.

I am deeply grateful to my colleagues and students, especially in the uniquely supportive and challenging intellectual community that is the Berkeley Department of Geography. Dick Walker, Michael Watts, Gillian Hart, Nancy Peluso, Lynn Huntsinger, and Louise Fortmann were wise mentors when I arrived and generous friends throughout; Kurt Cuffey, John Chiang, Rob Rhew, Jeff Chambers, and Laurel Larsen have embraced the challenge of sustained dialogue between the social and biophysical sciences with rigor and humor; Jake Kosek, David O'Sullivan, You-tien Hsing, and Paul Groth have helped me in ways large and small. James Bartolome has shared his wisdom about rangelands many times, not least in reading the penultimate manuscript version of this book in its entirety. Our cartographers, Darin Jensen and Alicia Cowart, worked to deadlines and details to produce the book's maps and line drawings, with assistance from Norm Miller, Sol Kim, and Xander Lenc. Nick Miroff and Adam Romero provided excellent research assistance. Finally, I am grateful to have graduate students—too numerous to list—who teach me as much or more than I can possibly teach them.

Without great archives and archivists, this book would not have been possible. I am grateful to the staff at the National Archives and Records Administration in College Park, Maryland—in particular, Joseph Schwartz and Peter Brauer—and in Denver, for assistance both while I was in the archives and in follow-up inquiries. The staff at Yale's Sterling Library and a string of researchers with the Rocky Mountain Research Station—in particular, Bryce Richardson—helped me track down key materials related to James Jardine and Arthur Sampson. Peter Hanff at the Bancroft Library here at Berkeley helped me access valuable archival materials as well.

Christie Henry and her staff at the University of Chicago Press have been both patient and helpful, not least in securing extremely constructive reviews of the prospectus and manuscript for this project. I am indebted to those reviewers, including Mark Stafford Smith, Curt Meine, and three who remain anonymous to me. I also want to thank David Carle and Stephen Pyne for responding to my inquiries about the history of fire research; Matt Turner for valuable suggestions related to chapter 6; Mark Westoby, Brian Walker, Ian Scoones, and Jeremy Swift for interviews and e-mail exchanges related to chapter 7; and Roy Behnke and Stephen Sandford for their assistance, also with chapter 7. A number of journal editors, including David Briske, Gary Meffe, Karl Zimmerer, and Mrill Ingram, have pushed me to publish, and to think and write more clearly, about these topics over the years. An earlier version of the second half of chapter 1 appeared as "The

Coyote-Proof Pasture Experiment: How Fences Replaced Predators and Labor on US Rangelands" (*Progress in Physical Geography* 39 (2015), 576–93).

A handful of good friends have been especially valuable to me, intellectually and otherwise, throughout the years of this research. I want to thank Aaron Bobrow-Strain and Jeff Banister for wonderful (if sometimes infrequent) conversations about history, the borderlands, and how to think about scholarship, and Geoff Mann for his wisdom in deciphering all matters political economic. Rodrigo Sierra Corona has amazed me over and over again with his ability to discover the most surprising things and fascinating people in and around Janos, Chihuahua. And Brian Powell, Leza Carter, and their kids have put up with me more times than I can count in my travels through Tucson: thank you for everything.

This book is dedicated to the past and the future: to the memory of my father, a geographer at heart, who followed this research for a decade but passed away just as I was finally starting to write; and to my beautiful children, who make every day fresh and alive. As for the present, mine walks contentedly beside Sasha Gennet, whose love lifts me up and holds the lot of us together: thank you.

Berkeley
22 February 2016

INTRODUCTION

Rangelands, Science, and the Politics of Scale

Attempts to convey the significance of rangelands almost invariably begin by invoking their vast extent: somewhere between one-third and one-half of all the ice-free land on Earth, one and a half times the area of all forests and two and a half times greater than croplands.[1] These facts have apparently failed as arguments, however: the terms *range* and *rangelands* generally evoke blank stares, because most people aren't sure what *range* even means. Not that they lack for *ideas* about rangelands: these are the places where both *Homo sapiens* and domesticated agriculture are believed to have first emerged, places that have long served as touchstones for fundamental questions of ethics, property, land, and government. Stories and parables about rangelands figure centrally in the world's major religions (Corbett 2005, Davis 2016) as well as in two of the twentieth century's most influential academic papers: Ronald Coase's (1960) "The Problem of Social Cost" and Garrett Hardin's (1968) "The Tragedy of the Commons." But rangelands and the science of rangelands have remained remarkably invisible to the general public and scholars alike.

This book examines the history of efforts to understand rangelands scientifically, and it proposes an argument for the importance of rangelands that can also account for their obscurity. Rangelands are indeed vast, and their vastness matters, but to say that they are important simply because they are so large puts the case inside out. After all, they are also relatively unproductive (in both ecological and economic terms), sparsely populated, and inhospitable compared to other kinds of land—shouldn't that matter as well? Indeed it should, and it does. The significance of rangelands derives from the obstacles and resistance they have presented (and continue to present) to three major forces of the modern world: the nation-state, science, and capital. Rangelands are difficult to "see" and control politically, to understand and manage scientifically, and to turn to a reliable profit economically. Vastness contributes to all three of these difficulties, but it is their interacting dynamics—rather than any fixed biophysical characteristic of the lands themselves, including size—that are of greater interest and consequence. Rangelands are sites where the separate and combined efforts of capital, science, and the state meet their limits, not in any fixed sense but as part of ongoing processes of trying to overcome and extend those limits. (From this vantage point, the closest analogue to rangelands is neither croplands nor forests but the open oceans, and, indeed, fisheries science and range science have much more in common than one might expect.) One result is a generalized public indifference to these lands, as society's primary attentions are concentrated elsewhere; another is a degree of social ambiguity and environmental uncertainty that is fertile terrain for storytelling, as in the Bible, and theorizing, as in Coase's and Hardin's articles. It is precisely their manifold marginality that enables rangelands to defy and disrupt social forces that elsewhere seem so powerful, and thereby to illuminate core tendencies, contradictions, and limitations in modern ways of knowing, using, and governing land and people.

What Are Rangelands?

Rangelands encompass grasslands, savannas, tundra, steppe, prairies, shrublands, and deserts. They include all lands that are neither cultivated, forested, covered by ice, nor built over—which is to say that what unites rangelands is less what they are than what they are not. Etymologically, the English noun *range* dates to the late fifteenth century and derives from the Old French verb *renger*, which denoted the unrestricted or peripatetic movement of herders and their livestock across large, open areas. This sense of the term persists, but in the twentieth century, range or rangelands became a residual category, comprising everything (other than ice-covered lands) that doesn't fit into more specific types such as forest, urban, or cropland (Sayre et al. 2013). The vegetation on rangelands "is always dominated by natural plant communities rather than by sown pasture. . . . As such, rangelands do not comprise a distinct ecosystem" (Grice and Hodgkinson 2002, 2–3). They may be dry or moist, hot or cold, nearly barren or thickly clothed in vegetation. Grasses, forbs, shrubs, lichens, biological crusts, rocks, or bare soil may dominate the ground surface; trees may be present or absent (though not dominant); soils may be among the most or the least fertile on Earth.

The association of rangelands with livestock grazing is etymologically correct but anachronistic and inexact. Many areas of rangelands are not grazed (or are grazed only by wild animals), and many non-rangelands are grazed (e.g., pastures and postharvest crop stubble).[2] Although not all domesticated animals can live everywhere, there are few vegetated landscapes that cannot support any livestock at all. If grazing looks like a defining attribute of rangelands, then, it is less because of its presence there than because of its absence or insignificance on other types of land. Moreover, because rangelands is a residual category, specific places can cease to count as such. About 35 to 50 percent of the world's wetter, more fertile rangelands have been converted to crop agriculture (Millennium Ecosystem Assessment 2005), and where irrigation has been developed, arid and semiarid rangelands have also been converted. Such lands are no longer classified as rangelands, and the rangelands that remain have therefore become, on average, drier and less productive.[3] Other rangelands have become cities and suburbs.

What unifies rangelands is thus not so much a biophysical or "natural" characteristic as a social one: marginality. Just as they are residual in classification systems, rangelands are peripheral to other, more lucrative types of land, and the people who inhabit and use rangelands are often (although not always) socially marginalized as well. Environmental conditions mediate this marginality, but they do not determine it. Frequently subject to weak or uncertain land tenure arrangements and viewed as empty or underutilized, rangelands are imperiled worldwide by fragmentation, land use change, and degradation of various kinds;[4] they might best be understood as nonforested places where intensive economic activities have not (yet) taken root. Similarly, whether for commerce or subsistence, range livestock production happens beyond or at the edges of the mainstream economic geography of the modern world. Even ranchers in the United States must absorb significant opportunity costs (by not converting their private lands to more lucrative uses) and are effectively powerless price takers relative to the highly consolidated beef industry (Fowler and Gray 1988; Barkema, Drabenstott, and Novack 2001, Torell et al. 2005).

Science and Rangelands in the Nineteenth Century

Humans have raised livestock on rangelands for thousands of years, but a specifically *scientific* engagement with them only emerged in the mid-nineteenth century. Livestock production served as "the principal means whereby Europeans colonized and then exploited the natural resources of sub-Saharan Africa, Australia, North and South America" (Grice and Hodgkinson 2002, 2), and the initial waves of commercial growth were soon followed by real or perceived environmental degradation, especially where domestic livestock had previously been absent (Milchunas and Lauenroth 1993). As environmental historians have shown, settler governments turned to science to reverse the damage and improve conditions for profitable production. Libby Robin (1997) characterizes ecology as a "science of empire," born to facilitate the extraction of agricultural commodities from colonial hinterlands in which rangelands figured prominently. British colonial officials mounted efforts along these lines in Australia, New Zealand, and sub-Saharan Africa after about 1840, and French administrators attributed rangeland degradation in North Africa to Arab pastoralists as early as the 1830s (Griffiths and Robin 1997; Holland, O'Connor, and Wearing 2002; Davis 2007). None of these European efforts gave rise to an institutionalized scientific field, however, instead remaining fragmentary and subordinate to other colonial endeavors such as forestry and crop agriculture.

In the United States, where the West served as a de facto colony of the East, rangelands were an enigma to explorers, settlers, and the government of the expanding nation throughout the nineteenth century. With their vast plains of grasses as tall as a horse, midwestern prairies left many European Americans bewildered and disoriented—some likened it to being lost at sea, unable to find their bearings (Kinsey, Roberts, and Sayre 1996). Others saw the region as a desert: Upon completing the Louisiana Purchase in 1803, President Thomas Jefferson referred to "immense and trackless deserts" in much of the area, and the explorers Zebulon Pike and Stephen Long subsequently confirmed the designation. Without trees suitable for timber, or much in the way of surface water, the area from the Missouri River west to the Rocky Mountains appeared "uninhabitable for people depending on agriculture for their subsistence," as geographer Edwin James wrote for the report of the Long expedition in 1823, and the accompanying map famously labeled the area "The Great American Desert" (James 1823). For the rest of the century, it remained unclear how these lands—let alone the even drier areas farther south and west—could be made to fit within established forms of land use, ownership, settlement, and law. Railroads and irrigation made commercial agriculture possible in some places, and mining booms drew people and capital into remote locations in the region's mountains and deserts. But huge areas remained unclaimed under the various Homestead Acts passed to encourage settlement from 1862 on, provoking political debates that burned through the 1940s and that flare up even to this day.

Like the British and French in their colonies, the US government sponsored scientific expeditions and research to study the West's landscapes and figure out how best to exploit them. But rangelands and livestock production were afterthoughts compared to minerals, railroad routes, and crop agriculture. John Wesley Powell's 1878 *Report on the Lands of the Arid Region of the United States* recognized the need for livestock raisers to control much larger areas of land than were allowed under the Homestead Acts, but irrigation and farming settlements were his principal concerns. Famous now for its prescience, Powell's report was summarily dismissed in Washington (Stegner 1954). More influential were the Public Lands Commissions of 1879 and 1903. Powell himself served on the first, which classified the public lands of the West according to their economic values: mineral lands had the highest value-to-area ratio, irrigable lands somewhat less, followed by timber lands. The least-valuable remainder were termed "pasturage lands," a precursor to today's rangelands. In the commission's judgment, only pasturage lands could be put to economic use without significant amounts of "aggregated capital," meaning that ordinary citizens might be able to settle them on their own (Williamson et al. 1880, xix–xxii). But almost nothing was yet known about the key resource for such purposes: grasses and other forage plants. By the time of the second Public Lands Commission, convened by President Theodore Roosevelt, the major forage plants had at best been described, and the nation's premier botanist, Frederick Coville, was among the commissioners. The commission's report (Richards et al. 1905) created the blueprint for the system of administration that persists to this day on some 300 million acres of western public rangelands, and Coville subsequently oversaw the research from which range science would arise.

The earliest scholarly journal article that might now be considered range science may be "Winter Grazing in the Rocky Mountains," published in the *Journal of the American Geographical Society of New York* by General Benjamin Alvord in 1883. Alvord was a mathematician, natural historian, and decorated Union veteran who had served in various parts of the West between 1852 and 1871. Originally delivered as an address, his article began:

> It is esteemed a proper subject for consideration and record in the American Geographical Society, that during the last fourteen years a revelation has dawned on the people of the United States, respecting the resources for winter grazing in the whole Rocky Mountain region. It is now known that all land over about 3,000 feet above the level of the sea has those qualities, viz.: that, without shelter, all the domestic animals can find ample food on the nutritious, summer-cured grasses of those plateaus, and that myriads of those animals are yearly raised by the great capitalists and others in our western regions. (Alvord 1883, 257)

Alvord sought to bring what he called "the philosophy of the cured grasses" to the awareness of the scholarly community. It had been learned almost by accident, he explained, when contractors building a telegraph line during the war had left their animals out to graze over the winter and found them healthy and fat the following spring. He devoted most of the article to an explanation of how climate, elevation, and forage plants made this possible.

Alvord's article had no discernible influence on subsequent range science, but it did document something important: "great capitalists" did not need scientists to inform them of the opportunity to profit from rangelands. Between the two great economic depressions of the nineteenth century, in 1873 and 1893, capital from the East Coast and the United Kingdom flooded across western rangelands, seeking profitable investment opportunities to absorb the surpluses produced by industrialism and empire (White 1991). Range livestock production was promoted as a spontaneous, natural source of wealth, in which plants and animals grew and reproduced by themselves. In the early years, it was famously easy: the bison had recently been eliminated, Native Americans had been conquered and confined to reservations, use of the land was free, and newly arrived cattle thrived virtually untended on native grasses, as Alvord reported. Boosters promised 20 percent returns on investment, and in some cases they weren't exaggerating. But profitability from livestock would become an ever more elusive prospect on US rangelands. The open range "beef bonanza" (Brisbin 1881) soon gave way to the paradigmatic case of "the tragedy of the commons," with starving livestock, bankrupt ranchers, and widespread severe overgrazing. Winter storms killed tens of thousands of cattle in the Great Plains in the 1880s, and severe droughts killed tens of thousands more in the Southwest in the early 1890s. The destructiveness of the droughts—vast areas reduced to bare dirt and cattle bones piled high for conversion into fertilizer (Bahre and Shelton 1996)—prompted the 1895 creation of the Division of Agrostology within the US Department of Agriculture (USDA) to conduct "grass and forage plant investigations."

Because this point of origin would have enduring and often overlooked implications for the history of range science, it is important to specify the mechanisms by which the cattle boom overwhelmed western rangelands. One of the earliest government scientists tasked with diagnosing the resulting catastrophe, H. L. Bentley, captured the dynamic eloquently:

> Every man was seized with the desire to make the most that was possible out of his opportunities while they lasted. He reasoned that there was more grass than his own cows could possibly eat. There was plenty of stock water for five

> times as many cows as were now on the range. There was no rent to pay, and not much in the way of taxes, and while these conditions lasted every stockman thought it well to avail himself of them. Therefore all bought cows to the full extent of their credit on a rising market and at high rates of interest. (Bentley 1898, 8)

The key point lies in the words *to the full extent of their credit.* Abundant credit both enabled cattle owners to expand their herds quickly and constrained them not to sell when prices crashed, forcing them instead to push on until every accessible patch of grass had been found and consumed (Abruzzi 1995). The root cause of the ecological destructiveness of the cattle boom, then, was a mismatch between the ecological variability of the West's rangelands and the international scale and invariable demands of capital flowing into the region. Range science was summoned into existence by and within the state to contain and remedy this crisis. Its task, from the outset, was to study rangelands with an eye to restoring their economic viability for capital, and to produce knowledge that could inform administrative policies and management practices that would serve this purpose. It was an early example of a "crisis discipline," in which scientists would be asked, "to make decisions or recommendations . . . before [they were] completely comfortable with the theoretical and empirical bases of the analysis" (Soulé 1985, 727).

At the end of the nineteenth century, however, the USDA and its Division of Agrostology had no authority to affect the West's rangelands, some of them privately owned but most still controlled by the Interior Department's General Land Office and governed by its mandate of disposal and settlement. Without a land base, none of the USDA's various bureaus "could do more than advise and research" (Pyne 1982, 191). The Forest Reserve Act of 1891 had authorized the president to withdraw timbered lands from disposal, however, and as the forest reserves grew in size and number, Congress and the Department of the Interior struggled to decide how to administer and manage them (Rowley 1985). The reserves were justified legally as a means to protect timber and watersheds, but they also encompassed large areas where livestock owners had been grazing their animals for decades or longer, and both forests and rangelands were considered to be in crisis due to unrestrained commercial exploitation. Forestry was a small but recognized scientific field, imported from Europe and first institutionalized at Cornell and Yale in 1898 and 1900, respectively; the chief of the USDA's Division of Forestry, Bernhard Fernow, could point to thousands of pages of published forestry research when he stepped down in 1898 (Steen 1998, 4). Research on rangelands, by contrast, had barely begun: little more than taxonomic and reconnaissance surveys had been completed by 1900 (Shear 1901), when the Division of Agrostology was combined with five other divisions into the Office of Plant Industry, renamed the following year as the Bureau of Plant Industry (BPI).

Then, in 1905, Congress transferred the forest reserves to the USDA and its new US Forest Service, headed by Gifford Pinchot and facilitated by his close friendship with President Roosevelt. As Pyne (1982) notes, the Transfer Act catapulted the Forest Service to the forefront of American conservation not on the basis of its scientific credentials—which were quite meager in comparison to, for example, the US Geological Survey—but by virtue of the 63 million acres of land that it suddenly controlled. For the new agency, the relationships between forestry and range, trees and grasses, timber and forage were at once scientific, bureaucratic, and management challenges. Convinced that science was the key to harmonizing conservation and utilization, Pinchot moved quickly to build research capacity, and by 1915 the Forest Service supplanted the BPI as the institutional home in which range science would develop.

Twentieth-Century US Rangelands Scholarship

The history of range science, especially its formative years in the United States, has been almost entirely overlooked by scholars. Apart from a single unpublished dissertation (Heyboer 1992) and one chapter of another (Pearce 2014), William Rowley's (1999) book chapter is the deepest treatment that historians have offered of the subject.[5] Otherwise the literature consists of a handful of articles by range scientists and administrators themselves,[6] the best by far being Chapline et al.'s (1944) valuable but dated article in *Agricultural History*. Many fine histories have been written about the government agencies that manage western rangelands, including the US Forest Service, the Bureau of Indian Affairs, and the Bureau of Land Management (BLM) (Peffer 1951; Foss 1960b; Steen 1976; Hirt 1994; Langston 1995; Merrill 2002; Weisiger 2009), but in most cases rangelands have been peripheral to other topics, and in all cases range science appears only fleetingly, if at all. Harold Steen's (1998) monograph, *Forest Service Research*, is almost wholly devoted to forestry, and even studies focused specifically on public range livestock grazing have said little or nothing about the role of science (Calef 1960; Culhane 1981; Rowley 1985). Similarly, scholars of the history of grassland ecology have said relatively little about range science or about rangelands outside of the Great Plains (Malin 1956; Tobey 1981). Finally, ranching has been the subject of many excellent historical and geographical treatments (Osgood 1929; Frink, Jackson, and Spring 1956; Atherton 1961; Jordan 1993; Remley 1993; Starrs 1998), but range science is virtually absent from them. Ranchers and government land agencies routinely feud over range management in most of this literature, but the scientific knowledge claims that informed the agencies' positions are not explored.

There is, of course, a much larger literature about historical events that took place in the West's rangelands: Indian wars, the slaughter of the bison, cattle barons and cowboys, stagecoach robberies, and shoot-outs with vigilantes and outlaws (Limerick 1987; White 1991; Isenberg 2000; DeLay 2008). Sometimes the land itself is just a passive backdrop, but in many cases its characteristics assume greater significance. Scholars of Western movies and novels, for example, have emphasized the dramatic aesthetic qualities of the West's vast open landscapes (Cawelti 1984; Slotkin 1985), and historians of westward expansion have accorded a central role to the region's aridity (Webb 1931; Worster 1979). Overcoming aridity thus becomes central; Stegner (1954), Reisner (1986), and Worster (1985), among others, have chronicled the history of irrigation development in ways that echo the narratives of Hollywood Westerns: engineers and government agencies operating at the edges of the law to subdue nature and secure the conditions for settlement and profit. But the scientists who endeavored to understand the ecology of the West's rangelands have almost completely eluded the attention of environmental historians (Alagona 2008; Isenberg 2014). As we will see, variability, even more than aridity, would prove to be the critical climatic factor.

US Rangelands Today

The chapters that follow describe in detail how difficult it was for scientists to produce the knowledge that we now have about rangelands, and specifically how prone the process was to faulty assumptions traceable to circumstances outside of the practices of science itself. This is not to say, however, that the efforts of range scientists have all been for naught. To assist the reader in situating the arguments of the book, the text here summarizes the state of our present knowledge about five major types of rangelands in the United States; I place it in a box to signal a suspension of the book's larger epistemological lens. Much of the material is drawn from Holechek, Pieper, and Herbel (2004).

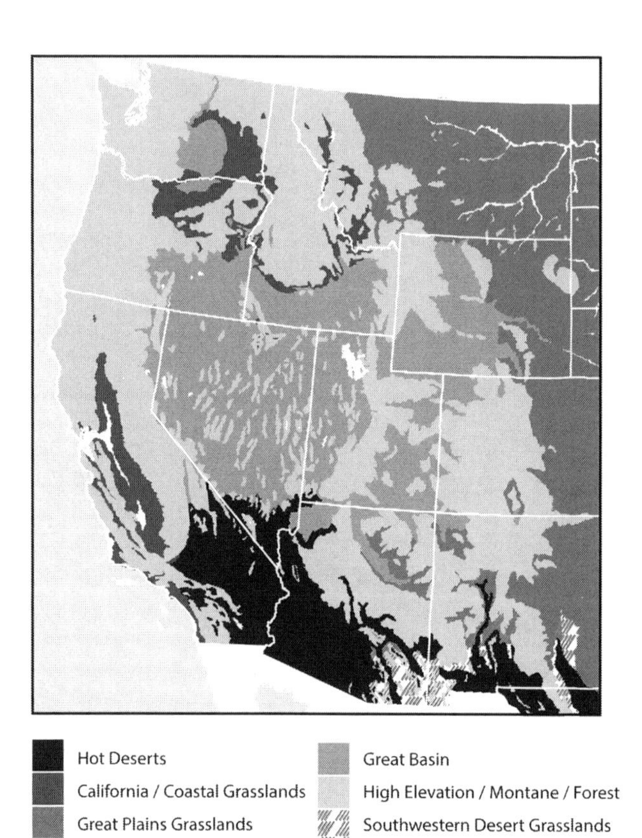

Map I.1. Major rangeland types of the western United States.
Adapted from USDA (1936) by Darin Jensen.

The rangelands of the western United States are diverse. They can be classified for our purposes into the five types shown on map I.1: Great Plains Grasslands, Hot Deserts, California/Coastal Grasslands, Great Basin, and Southwestern Desert Grasslands. (The sixth category shown on the map, High Elevation/Montane/Forest, contains some areas of rangelands, but it is primarily forest or, at the highest elevations, rock and ice.)

Great Plains Grasslands. The Great Plains extend from Canada nearly to Mexico and from the Rocky Mountains eastward, grading into the Midwestern prairies that once reached as far as present-day Indiana. Grasses and grass-like plants predominate, completely covering the ground surface and forming a dense sod. Sloping gently downward toward the Mississippi River, the Great Plains and the prairies—shown in map I.1 as a single type—are distinguished first by a difference in rainfall: the lower, wetter, eastern portion is known as tall-grass prairie, while the higher, drier, western portion is dominated by short grasses. The line between them falls roughly along the 100th meridian, aligning with the twenty-inch isohyet and famously known as the western limit of reliably feasible farming without irrigation. A third zone of mid-grass (or mixed) prairie is often demarcated around this line—this was where Frederic Clements did his pioneering ecological work, about which we will have much more to say. By carefully measuring vegetation in meter-square plots across the length of Nebraska, Clements was able to show that the mid-grass prairie moved, so to speak: taller grasses dominated during wetter years, shorter ones in drier years, each group of plant species responding to the transient conditions that favored it more reliably to the east and west, respectively (Pound and Clements 1898). The relative proportions of short-, mid-, and tall-grass species in the vegetation changed gradually, like the average rainfall. Clements derived his theory of plant community succession, in part, by converting this spatial gradient into a temporal one (Tobey 1981).

The key points about the Great Plains are four. First, for thousands of years they were home to some 60–100 million bison and other large grazers, meaning that the vegetation coevolved with grazing and was adapted by evolutionary selection to withstand it. Domestic livestock such as cattle therefore found a preexisting ecological niche there. Second, rainfall in the Great Plains, although variable, is greater than in most other rangeland areas of the United States: average annual precipitation in Nebraska, for example, grades from about fifteen inches in the west to about thirty inches in the east. Forage production is correspondingly higher and less variable than in the other major rangeland types (table I.1). Third, while most of the drier, short-grass prairies near the Rocky Mountains (where average rainfall is closer to ten inches annually) remain today, more than 96 percent of the tall-grass prairie and more than three-fifths of the mid-grass prairie have been converted to croplands (Samson, Knopf, and Ostlie 1998). Part of Clements's motivation for his work was to learn as much as

Table I.1. Present extent, productivity, and ownership of seven major rangeland types in the United States

Rangeland type	Forage production (kilograms per hectare dry matter)	Area (million hectares)	Percent in federal ownership
Tall-grass prairie	1,500–3,500	15	1
Mid-grass prairie	1,000–2,500	50	17
Short-grass prairie	600–1,000	20	5
California annual grassland	400–3,500	3	6
Palouse prairie	300–1,000	3	15
Great Basin (cold desert)	100–500	73	77
Hot deserts	150–500	26	55

Adapted from Holechek, Pieper, and Herbel 2004, 87.

possible about the prairie ecosystem before it disappeared under the plow. Fourth, the theory Clements developed in the Great Plains went on to be applied to all rangelands, notwithstanding these differences.

Hot Deserts. Three hot deserts are merged in map I.1: the Mojave in California and parts of Arizona and Nevada, the Sonoran in southern Arizona, and the Chihuahuan in Texas, New Mexico, and southeastern Arizona. The Mojave is the driest of the three, with only four to eight inches of precipitation annually; it is primarily a shrubland. Shrubs and cacti predominate in the Sonoran desert, which is similarly hot but more diverse than the Mojave due to slightly higher rainfall (two to twelve inches annually), especially in summer, and the near absence of winter frost. The Chihuahuan desert receives six to twelve inches of precipitation, mostly in the summer, and is less diverse due to more frequent frost. In all three deserts, forage plants are relatively scarce except for brief flushes of annual grasses and forbs following heavy rains; bare ground often predominates, and livestock grazing is minimal or nonexistent, although it occurred in much of the area in the late nineteenth to early twentieth centuries.

California/Coastal Grasslands. This category combines California annual grasslands and Palouse prairie (shown separately in Table I.1), as both were once dominated by bunchgrasses. The Palouse prairie—located in eastern Washington and Oregon and in small patches of Idaho and Montana—has largely been converted to wheat production. Large parts of the California annual grasslands also are now croplands, especially in the San Joaquin, Sacramento, and Salinas valleys and along the southern coast (where many have in turn been urbanized). They persist where irrigation has not been developed, supporting significant herds of livestock. As shown in table I.1, the California annual grasslands are the most variable of all western rangelands in terms of forage production. The Mediterranean climate predisposed them to early and thorough invasion by annual grasses from Europe with the arrival of the Spanish, and their vegetation dynamics are unique among US rangelands (Perevolotsky and Seligman 1998). California annual grasslands do not play a significant role in the history told here, primarily because of the very minor presence of the Forest Service (and the federal government more generally) in their ownership. The study of these lands by range scientists is well developed, but it occurred slightly later and was led by scientists from the University of California, institutionally separated from the mainstream of the discipline (Stromberg, Corbin, and D'Antonio 2007).

Great Basin. Comprising almost all of Nevada, large parts of Utah, Wyoming, Idaho, Oregon, and Washington, and portions of western Colorado, northern Arizona, and northern New Mexico, Great Basin rangelands as depicted on map I.1 include sagebrush grasslands and salt desert shrublands. Sagebrush grasslands, sometimes known as shrub steppe, make up the majority of the area, and are higher and wetter (eight to twenty inches, with an average of about ten) than the salt desert shrublands (three to ten inches, with an average of five). More than three-quarters of the Great Basin is federally owned, but the military and the Bureau of Land Management are more prominent than the Forest Service, which probably explains why significant range science effort was not invested there until the 1930s. Much of that effort, when it occurred, focused on converting shrub steppe by killing sagebrush and seeding imported grasses (chapter 5).

Southwestern Desert Grasslands. This type—the smallest in area of the five types shown on map I.1—occupies the transition zone around the continental divide between the Sonoran and Chihuahuan deserts in southeastern Arizona, southwestern New Mexico, and far western Texas. Rainfall is slightly greater than in the adjacent deserts, but highly variable in both time and space, with annual averages ranging from about eight to twenty inches depending on elevation. When livestock first arrived in this area with the Spanish explorer Francisco de Coronado in 1540, the vegetation was dominated by perennial bunchgrasses, many of them related to those of the short-grass prairies. Some evidence of bison has been found in the area, but large grazing animals were not numerous in recent evolutionary time. The bunchgrasses were highly palatable to livestock, and when droughts interrupted the cattle boom, extreme overgrazing had lasting effects. Today, shrubs and woody plants have replaced perennial grasses across large parts of the region. The combination of great forage production followed by catastrophic collapse motivated the creation of the two earliest and largest federal range research stations in the Southwestern Desert Grasslands (chapter 3), which in turn helped give this type of rangeland disproportionate importance to the history of range science.

Summary of the Argument

Rangelands vary so widely that biophysical generalizations about them are difficult to sustain, and the challenge of encompassing them within a single scientific field of knowledge was attended by considerable difficulties. To make rangelands a coherent object of scientific apprehension involved observing, describing, classifying, and analyzing the climates, soils, plants, and other organisms found there; organizing the information internally and in relation to knowledge from other lands and relevant scientific fields; and developing theories from which to formulate hypotheses, experiments, predictions, and management recommendations. This ambition to systematic, rational-logical, predictive, and "objective" knowledge is what distinguishes the efforts examined in this book from other ways of knowing rangelands. Practical, experiential, or craft knowledge can be extremely sophisticated and deep (Meuret and Provenza 2014), but it is not my focus here because range science discounted or ignored it until very recently.

It has long been agreed that for most of its history, range science rested on an ecological theory associated with Frederic Clements (1874–1945) and his two-volume magnum opus, *Plant Succession* and *Plant Indicators* (Clements

1916, 1920). More recently, it is also agreed that this foundation was a mistake: Clementsian, or successional, theory is now seen as ill-suited to many rangelands, especially in arid and semiarid settings (Box 1992; National Research Council 1994). It works in "equilibrial" ecosystems such as the Great Plains, where Clements did his pioneering work as a student and then as a professor at the University of Nebraska between 1892 and 1907. But in "nonequilibrial" ecosystems, characterized by less precipitation and/or more variable precipitation than in Nebraska, successional theory cannot explain the dynamics of vegetation, with or without the influence of livestock. So-called nonequilibrium theory did not gain traction in range science until the 1980s, however—more than half a century after the discipline embraced succession. How did range science make this fateful mistake, and why did it take so long for the discipline to recognize it? Other missteps, it turns out, are related to this one. Why was the policy of fire suppression, instituted by the Forest Service with increasing vehemence and effectiveness after 1910, applied not only to forests but to rangelands, even where there was little or no marketable timber? With the benefit of hindsight and archives, I show that the conventional answers to these questions are incomplete or inaccurate, and that these mistakes could have been avoided: early range scientists recognized them and attempted to voice their objections, but the institutional contexts in which they worked overrode them.

Chapter 1 examines extermination and fencing, which came together in the early twentieth century to produce "the range": the sociospatial order that range science would take for granted for decades to come. Campaigns to exterminate "noxious" wildlife—principally rodents and predators—had been initiated by European American settlers as they spread across the West. State and county governments often encouraged them with bounty payments. When the resulting patchwork of bounty schemes proved ineffective and costly, a federal role emerged, first for research and extension to assist landowners and then for directly executing extermination campaigns with or without landowner cooperation. The USDA's Bureau of Biological Survey led these efforts under a 1914 statute, killing hundreds of thousands of predators and hundreds of millions of small mammals across vast areas of rangelands. Two of the most drastically reduced animals—prairie dogs and wolves—are now considered keystone species, meaning they may have played disproportionately important ecological roles in the landscapes they inhabited. But their absence was little noted by range scientists who sought to see or imagine the original, pre-cattle boom conditions of the lands they studied, because extermination was a fait accompli carried out by an agency other than their own.

Predator control and fencing were the subjects of the single most influential experiment in the history of US range science: the Coyote-Proof Pasture Experiment, conducted in the Wallowa National Forest in eastern Oregon between 1907 and 1909. The goal of the experiment was to see whether a herd of sheep would grow and produce wool more efficiently in a space where predators were absent and human herders (and their dogs) were therefore unnecessary. It was declared a success almost before any data had been collected, and the final interpretations were seriously flawed, but the experiment prompted the Forest Service to take over range research from the BPI and to install the two young men who led the experiment, James Jardine and Arthur Sampson, in positions from which they would profoundly shape the nation's range administration and research, respectively. US rangelands would henceforth be fenced as the first step both for management—to allocate access and minimize the costs of labor—and for scientific research—to control and measure stocking rates (the number of livestock relative to an area of land). Predator-proof fences were prohibitively expensive, however, so eliminating herders—the underlying goal of the experiment—would require continuous wholesale extermination of predators. The near-absence of predators and the ubiquity of fences, along with exclusive land tenure, thus became unacknowledged assumptions or blind spots for range science.

Chapter 2 explains the origins of another blind spot, fire, and its relation to Clements's theory that succession was a universal tendency of all plant communities. In keeping with widespread assumptions about the natural world, Clements conceived of vegetation as in balance or equilibrium with the climate and soils found in any given location, unless disturbed by some agent or event such as drought, fire, heavy grazing, or human activities such as plowing. Following disturbance, the same processes that produced a given vegetation community over very long time scales (so-called primary succession) would drive a return to the original or "climax" equilibrium conditions (secondary succession). Clements recognized the role of recurrent burning in many grasslands, but he considered it to be an external disturbance that interfered with succession, resulting in a "fire disclimax" rather than a true climax. For the Forest Service, meanwhile, all fires were unequivocally bad, not only for timber but also for watersheds, forage, and wildlife as well as people and property. Until about 1930, the relation of livestock grazing to forest fires was recognized and discussed, at first openly and then only internally within the Forest Service. There was strong pressure on and in the agency to graze the national forests heavily as a means of preventing and limiting fires, but a policy of "overgrazing" was impolitic.

Where was the line between too little and too much grazing? Despite their best efforts, range scientists could not answer this question definitively.

It was in this context that Sampson, who had studied under Clements at Nebraska, published a famous 1919 Forest Service Bulletin entitled *Plant Succession in Relation to Range Management* Sampson proposed that overgrazing could be detected early by studying plant composition for signs of departure from climax conditions: grazing was counter-successional, and stocking rates therefore could be set to balance successional forces and maintain the vegetation in a desired state. His bulletin is widely cited as the moment when successional theory became the foundation of range science. But it was administrators, not scientists, who were actually responsible, and it did not really happen for another decade. Jardine misinterpreted Sampson's argument to mean that once they were set correctly, stocking rates could be static, avoiding the difficulty of persuading or forcing ranchers to adjust their herds to fluctuating rainfall and forage production. In the late 1920s, as part of an administrative reorganization and a significant increase in funding, the head of the Forest Service's Branch of Research, Earle Clapp, declared that successional theory would guide all range research, even while he acknowledged that the theory had barely been tested. Circumstantial evidence suggests that this decision was silently motivated by the goal of fire suppression and enabled by Jardine's misinterpretation: static stocking rates ensured that there would be little or no grass to fuel fires during severe dry years, when the risks were greatest.

Chapter 3 considers these events from the perspective of scientists working at the two largest and oldest federal range research stations: the Santa Rita Experimental Range in Arizona and the Jornada Experimental Range in New Mexico. Founded by the BPI in 1903 and 1912, respectively, they were taken over by the Forest Service in 1915 to serve as sites for experimental research and for demonstrating the benefits of scientific management to ranchers in the region. Scientists at both locations struggled to make sense of the dynamics of vegetation and grazing in the face of rainfall patterns that were unpredictable and highly variable in both space and time. Many of their observations pointed away from Clements's model of succession and climax; contrary to Sampson, the weather, not grazing, seemed to be the dominant driver of vegetation change. These scientists rarely used Clementsian concepts in their reports, publications, and internal communications until the 1930s. By that time, another problem had emerged: even in areas without any livestock, former grasslands were turning into shrublands dominated by mesquite and other woody plant species. Southwestern range scientists pointed out that succession was not working as predicted, but they lacked an alternative theory to replace it.

The consolidation of successional theory as the foundation of range science profoundly shaped the management and politics of US rangelands as well as research. It meant that "original" conditions would serve both as the touchstone for understanding rangeland vegetation and as the presumptive goal of management. This occluded the possibility that some changes might be effectively permanent (or at least not reversible simply by removal of the disturbances that caused them), or that the same site might stabilize in different states depending on the timing or sequence of quasi-random events such as extreme droughts, wet spells, fires, or killing frosts. It also blocked from view the roles that Native Americans or other now-missing forces (e.g., fire, beavers, large predators, or prairie dogs) may have played in shaping presettlement conditions (Dobyns 1981). Furthermore, Sampson's adaptation of Clementsian theory conformed to, and ratified, the key concern of administrators and ranchers alike: stocking rates. Bureaucrats saw them as the primary administrative variable, and ranchers (and their creditors) saw them as the primary economic variable. Scientists, henceforth, would treat them as the primary ecological variable. Environmentalists embraced the same ideas when they joined the debates later in the century, convinced that if fewer livestock were better for rangelands ecologically, then no livestock would be best of all. It should not be surprising that these debates persisted for so long without resolution because in most western rangelands the underlying theory was flawed.

Chapter 4 looks at the scientific struggle to determine carrying capacities and the political struggle to impose them on ranchers in the form of stocking rates. It took the Forest Service decades of concerted effort just to develop a basic inventory of its rangelands, and even when data were available, the problem of converting an ever-changing volume of forage into a fixed number of livestock was a matter of chronic uncertainty and debate. Fixing stocking rates to match average conditions ensured that there would be excess grass in wet years and excessive livestock in dry ones; perennial disputes with ranchers predictably ensued. Publicly, the agency made bold claims about the condition of the nation's rangelands, invoking solid-looking numbers to defend itself against ranchers, rival agencies, and critics in Congress. Internally, however, the scientists conceded that determining carrying capacities was fraught with problems. Then, at midcentury, an apparent solution was found: economical, standardized methods of quantifying *both* the successional stage *and* the volume of forage of a given range. It was not actually a solution, as would become apparent over time, but it represented an important final step in consolidating the authority of range science.

Meanwhile, rangelands in large parts of the West continued to defy the expectations of Clementsian theory. Chapter 5 examines efforts to counter the real or perceived "invasion" of brush and shrubs by artificial means when "natural" improvement by succession failed to occur. From the days of the Division of Agrostology, scientists had investigated seeding and cultivating rangelands using both native and imported, wild and domesticated grasses. They had also tried various methods of killing shrubs. But western rangelands were simply too big, relative to the amount of profit they could produce, to justify the expense of intensive treatments, and cheaper methods repeatedly failed. After World War II, however, the prospect of artificial improvement found new life thanks to postwar prosperity, new technologies, and chemical herbicides. It also attracted political patronage as a way of defusing—or at least distracting attention from—the tensions between agencies and ranchers over stocking rates. Seeds were tested and mass produced, especially for several nonnative species; heavy equipment and tractor attachments were developed and deployed; and newly available herbicides were applied, often from airplanes. Huge areas of sagebrush rangelands in the Great Basin were forcibly converted to grasses. But several nonnative grasses became invasive species in their own right, and mesquites in the Southwest persisted. By the 1970s, public outcry over the use of toxic chemicals (in both the United States and Vietnam) combined with rising oil prices to dampen the dream of manipulating rangelands artificially. As faith in natural improvement by succession wanes, however, artificial measures show signs of resurgence under the banner of ecological restoration.

The last two chapters examine the overseas travels of range science in the postwar period. It had been an inward-looking, US-centric discipline throughout the first half of the twentieth century, but it was the only institutionalized science dedicated to rangelands at the end of World War II. Chapter 6 explores how range science was exported in international pastoral development programs. None of their internal debates and frustrations at home prevented range scientists from taking the stage at major international conferences to proclaim the value of their knowledge. They presented the Southwest not as evidence of their own failure but as proof of the urgency to intervene in other arid and semiarid rangelands to prevent "desertification"; outside capital was portrayed not as the cause of rangeland degradation but as the solution to it. The concept of carrying capacity lay at the heart of a rising neo-Malthusianism in public debates and foreign policy circles, and the imperative of controlling stocking rates merged seamlessly with calls to prevent overpopulation from driving the world's poor into the communists' arms. The World Bank, the United Nation's Food and Agriculture Organization, and other multilateral agencies took ranching and range science as the model of modern livestock production, and the sociospatial order and assumptions embedded in the discipline were imposed on pastoral communities and their rangelands in Africa, Asia, and Latin America.

By the late 1970s, however, it was clear that virtually all these pastoral development projects were abject failures. Chapter 7 describes how these failures, combined with a growing body of rangeland research overseas, finally dislodged Clementsian succession as the theoretical foundation of range science. European social scientists defended the economic rationality of subsistence pastoralists and challenged the "tragedy of the commons" thesis that communal land tenure led inevitably to overgrazing and environmental destruction. Australian ecologists began to study their rangelands and noticed that US range science couldn't explain what they were seeing. Through the International Biological Program, ecologists from Europe, Africa, Australia, and the Middle East encountered US rangelands and range science, exchanged ideas, and went back to their own rangelands searching for alternatives. Systems ecology and modeling provided them with ideas and tools that could account for the anomalies long observed on the Santa Rita and Jornada Experimental Ranges. By 1990, ideas from all these groups and places had come together into a nonequilibrium theory of rangeland ecology that is now seen as superior to Clementsian succession in most settings, especially where rainfall is limited and highly variable. Whether the new theory will succeed where the old one failed, however, remains to be seen.

Methods and Sources

The sources for my research were broadly of three kinds: the published scientific literature about rangelands; memos and other internal records found in various government archives; and my own observations and experiences with rangelands, range scientists, and ranchers over the past three decades. Each entailed certain methods—archives, ethnography, and interviews being the most prominent—and the outcome required triangulating among them. Each also had certain social forms that conditioned their production and therefore how I handled them epistemologically.

The published scientific literature mostly falls into two categories: scholarly journal articles and government reports—the latter usually not peer reviewed and therefore potentially different in important ways from the former.

Constituting what might be termed "official" knowledge about rangelands over time, this literature tends to be written in the present tense. Past errors are omitted entirely or mentioned only in passing, and information deemed extraneous or accidental to the formal requirements of publication—experiments that failed, findings that were anomalous or deviated from official policies, or the broader social conditions that shaped the research questions—is excluded or minimized.

Archival sources serve to remedy these shortcomings in the published literature. How scientific knowledge about rangelands was produced, including knowledge that was later considered mistaken, can be discerned from memos, notes, letters, internal reports, and other materials generated by the institutions in which range scientists worked. Especially in the period before 1950, most of these institutions were agencies of the US federal government, which, to the great good fortune of scholars, keeps meticulous track of them through the National Archives and Records Administration and in the archives of many range research stations. Although also constrained by formal requirements, these materials are less "official" and less public than the published literature, and in many cases, they reveal questions, disputes, rivalries, and doubts about the official knowledge, as well as the intra- and interagency relations that conditioned the production of that knowledge.

For both published and unpublished, official and unofficial sources, formal requirements or expectations were themselves historically conditioned and subject to change. Range-related government reports, for example, were more open-ended and qualitative in the 1890s and 1900s than they were in the latter half of the twentieth century, much as peer-reviewed articles tended to become more specialized and technical over the same period. The paperwork and protocols of government agency operations also changed over time, requiring and enabling different kinds of inferences and insights. And written memos, which put something on the record in a bureaucratic context, are, of course, only relatively private, although that presumably makes the expression of divergent or controversial views all the more noteworthy. In some respects, the changing forms of producing and presenting knowledge may be more historically significant than the particular content contained in most documents.

My own experiences with rangelands and range science provide a third source of information and understanding for the analysis in this book. I first lived and worked on rangelands between 1986 and 1989 in eastern California as a student at Deep Springs College, which operates a cattle ranch on irrigated private land and large grazing allotments leased from the US Forest

Service and the Bureau of Land Management. Later, working for the Arizona Conservation Corps, I built many miles of barbed-wire fence to keep livestock out of public lands owned by counties, the state, and various federal agencies; I also got to observe firsthand the struggles of government employees coping with bureaucratic rules and procedures that often seemed byzantine, impractical, or absurd. These experiences subsequently proved invaluable during my dissertation research, also in Arizona, conducting ethnographic fieldwork with ranchers and employees of the US Fish and Wildlife Service. Since 1996, I have interviewed scores of ranchers, reconstructing the history of their lands and management practices and often trying to untangle the thicket of regulatory issues that they must navigate, especially on public lands. I have also worked with groups of ranchers on community-based conservation efforts (Charnley, Sheridan, and Nabhan 2014) aimed at restoring fire and grasslands, protecting endangered species and habitat, and preventing subdivision of private rangelands into residential home sites. In 2001–2004, I held a postdoctoral fellowship with the USDA Agricultural Research Service's Jornada Experimental Range in southern New Mexico, and since that time I have maintained ongoing collaborations and friendships with range scientists there and elsewhere, in what might be considered participant observation, both in the field and at conferences and professional meetings. Although I have never been formally trained in range science or ecology, I have immersed myself in these literatures for nearly two decades. Taken together, these experiences have given me insights into the livelihoods and ways of thinking of the three main classes of actors relevant to this history: ranchers, range scientists, and government agencies.

Mirroring my own experiences, this book focuses disproportionately on the Southwest, especially the desert grasslands and shrublands of southern Arizona and New Mexico. The reasons for this are more general, however. The history of southwestern range science has been even more neglected in the literature than its history elsewhere. This is somewhat odd, because the earliest and most important range research stations were established in the Southwest, and until the 1930s they received the bulk of the federal government's expenditures for range science. It was also in the Southwest that several important techniques for surveying and monitoring rangelands were developed before being applied nationally. Despite all the research effort, however, southwestern rangelands have always been the most intractably difficult for scientists and agencies to understand and administer in line with national policies and procedures. Although they differed in important ways from other rangelands in the United States, they were better analogues for many rangelands overseas, which proved important in the last third of the twentieth century.

Coproduction and the Politics of Scale

This book is intended both for practitioners—scientists, ranch managers, environmentalists, and government employees involved with rangelands—and for scholars interested in geography, political ecology, science and technology studies, and environmental history. The epistemological approach is best described as critical realist (Sayer 1992), and, although I do not foreground issues of social theory, they nonetheless inform and undergird the empirical materials that drive the narrative. Proceeding historically, I show how scientific knowledge was *coproduced*, not only in the general sense that "knowledge and society continually shape each other" (Forsyth 2003, 104) but, more specifically, in that social forms and forces conditioned the very categories and practices of range science: they did not act *on* range science as though from outside of it, but rather were *inside* range science, conceptually and otherwise, from the very beginning. The concept of carrying capacity, for example, which appears without definition or discussion in the earliest reports of the Division of Agrostology, originated elsewhere altogether (Sayre 2008), and it better reflected the needs of bureaucrats and bankers than those of scientists or ranchers.

"The range" was *produced* as space and nature (Smith 1984), in minds and on the ground, for purposes of the state, capital, and science simultaneously: to territorialize vast, marginal landscapes for government administration, facilitate their profitable use, and enable experimental control and analysis. This sociospatial form ensured that rangelands would be governed, managed, and studied in the mode of ranching, based in social relations of exclusive land tenure, credit, and market exchange; other possibilities were excluded or effectively rendered invisible. Not that the logics of state territoriality and capital accumulation unfolded in smooth concert or without considerable friction: there were chronic struggles between and among government agencies, elected officials, and ranchers, and "scientific" management frequently failed to fulfill ecological and/or economic ambitions. Ideas about national identity, race, and ethnicity—applied to plants and animals as well as people—impinged in various ways, and nonhuman phenomena—organisms, populations, and the weather—frequently behaved in unexpected or inconsistent ways. The vast spatial extent of rangelands, often accompanied by remoteness, harsh climate, and rugged terrain, compounded the challenge:

how could scientists *see* landscapes that were so large, diverse, and inaccessible? If there had been greater economic opportunities at stake, government or industry might have stepped forward with more funding to support such extensive research. But on a per-unit-area basis, rangelands were generally the *least* lucrative places. Moreover, the temporal variability of many rangelands—a function of low and erratic rainfall—meant that a snapshot of their conditions at a single point in time could be highly misleading: scientists needed to see them repeatedly to measure them accurately, let alone understand them.

The value of science, in this context, derived from its putative apolitical objectivity and predictive power, enabled and ensured by the methods and practices of science itself. But range science could not fulfill these ascriptions. As a practical matter, the vast extent of western rangelands long exceeded what was financially, logistically, and technologically feasible in terms of data collection and research, and even when more resources were forthcoming, other problems remained. Moreover, whether accurate or erroneous, the scientists' findings could not avoid having political origins and consequences, because roughly half of US rangelands—some 300 million acres—were (and remain) publicly owned (map I.2). Political objectives varied over time, from settling the western United States with European Americans, to protecting or restoring timber, forage, and watershed values, to conserving endangered species and biological diversity. Meanwhile, increasing economic returns through livestock production, land appreciation and speculation, or other means was a persistent though elusive ambition for many bureaucrats, politicians, and scientists as well as ranchers. Range science could lend authority to agency positions, however, only by erasing any signs of extra- or nonscientific influences and assumptions; it had to appear to be what it was not—an accurate, predictive understanding of grazing and rangeland ecology—and *not* to be what it was—a body of knowledge summoned into existence to address fundamentally political and economic problems. The issue of scale was the pivot on which this occlusion and erasure turned.

The Politics of Scale provides an extended empirical case study to complement theoretical arguments I have made elsewhere (Sayre 2005a, 2009, 2015; Sayre and Di Vittorio 2009). I seek to integrate ecological (Levin 1992; Peterson and Parker 1998) and geographical (Cox 1998; Swyngedouw 1997; Brenner 2000) notions of scale and to illuminate how scale is always already political *and* scientific, epistemological *and* ontological, socially produced and historically contingent *and* biophysically conditioned. As suggested earlier, the severe and widespread degradation of rangelands in the

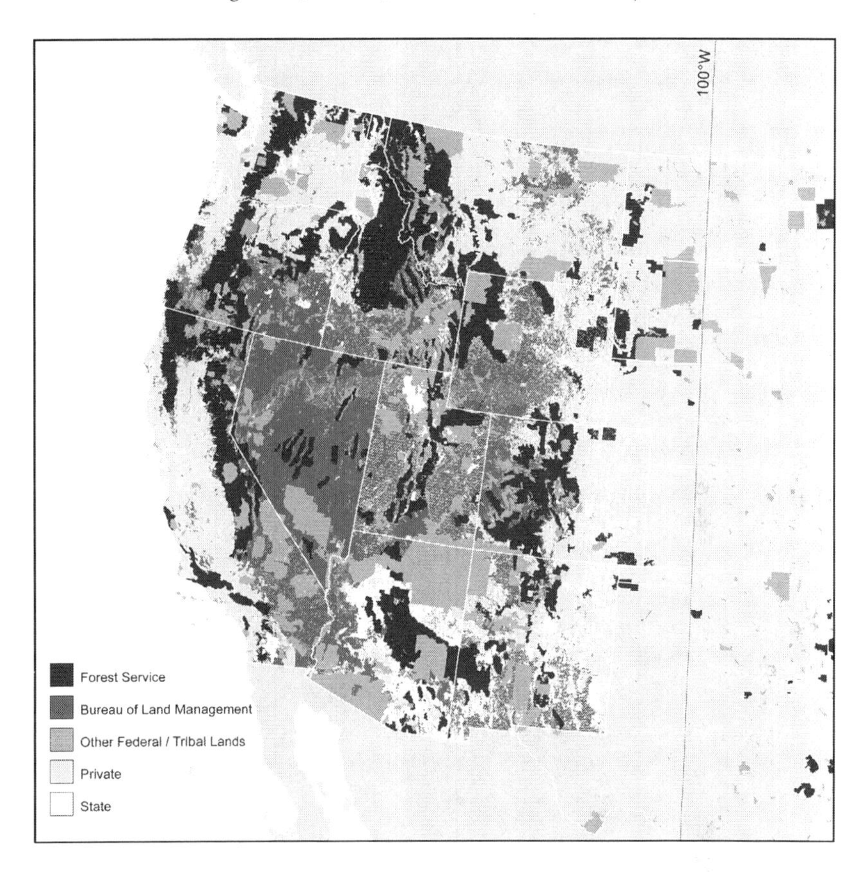

Map I.2. Landownership in the western United States. The federal government leases 265 million acres of its western rangelands to private livestock producers; another 41 million acres of other public lands (owned by states, counties, cities, etc.) are similarly leased. The total acreage in livestock production in the eleven western states is 525 million acres (Wuerthner and Matteson 2002, 5). By Darin Jensen and Alicia Cowart.

western US resulted from the scale of capital flows relative to rangeland ecosystem processes in the late nineteenth century. In response, the Forest Service and other federal agencies sought to impose and legitimate management and governance of rangelands at a national geographical scale. Scientists were enrolled to produce both practical and ideological tools for these efforts, and developing ecological scales at which to measure and interpret rangeland ecosystems was central to their work. The scientists' choices were constrained by economic and political forces, however, and the temporal grains of their measurements repeatedly proved too coarse. Estimating forage demand in animal unit months, for example, elided the dynamism of grass growth, which was better captured at daily rather than monthly resolution; similarly, annual averages of precipitation elided both the interannual variability of actual rainfall and the fact that the same amount of rain could have dramatically different effects depending on when and how it fell during the year. "Scales are a joint product of social and biophysical processes . . . constrained overtly by politics, and more subtly by choices of technologies, institutional designs, and measurements" (Lebel, Garden, and Imamura 2005, 1).

If the national scale of governance was poorly suited to rangeland ecosystems, the spatial and temporal scales of range science failed to match the scales at which its findings were applied. In the first edition of their highly influential textbook, *Range Management*, Laurence Stoddart and Arthur Smith (1943, 177) observed that "There are two major groups of methods, as follows: extensive, or general; and intensive, or detailed. The former, which hold a very definite place in range management, are used primarily in applied administration; the latter are used mostly in research studies." They did not elaborate on the problems this created, however. Most range *research* took place at the scale of plots or pastures, where experimental controls and intensive measurements were feasible; this was what made the findings scientifically credible and authoritative. But range *administration* took place at much larger scales: federal laws and policies governed the key public lands agencies, organized into regions or districts that encompassed multiple states. Meanwhile, range *management* typically took place at intermediate scales, much larger than experiments but much smaller than regions. The scale of each grazing allotment on public lands was large enough to present a major challenge even for basic inventories, and there were tens of thousands of allotments to measure. Averages across regions or national forests were not sufficient, because conditions varied at smaller scales and because each allotment was leased to a rancher whose livelihood depended on conditions in that specific allotment. Moreover, conditions could be measured in various ways, each with its own temporal scale: the *amount* of forage could vary rapidly from month to month and year to year, for example, while the *composition* of forage plants might change only more slowly. The first was conspicuous and was generally what ranchers considered important, whereas the second was harder to detect without technical measurements and scientific expertise—and according to Clementsian theory, it was the more important indicator of range condition.

In other words, the problem of scale was not simply a matter of mismatches, as though all that were needed was correct alignment of one scale to another. Rather, it concerned the relations between processes operating at different scales, and the need therefore to understand multiple scales simultaneously. Through the lens of Clementsian ecology, however, this problem was invisible. Quadrat methods of vegetation measurement, which Clements pioneered, provided precise quantitative data at small scales, and each quadrat could be identified with a plant community described in *Plant Indicators*, which exhaustively cataloged the vegetation of the entire West (Clements 1920). The relationship between smaller and larger scales was assumed to be linear: plot data were multiplied by area to calculate forage values for pastures, ranches, and larger areas. On the temporal axis, Clements's theory did not span scales so much as collapse them, by equating short-term recovery from disturbance with long-term succession to climax. It followed that no matter how degraded a site may have become, it would necessarily recover in a predictable way once the disturbance abated: "This," he confidently proclaimed, "is the basis of all range improvement" (Clements 1920, 310). For scientific purposes, this permitted results from a period of years to be averaged and extrapolated into the future, thereby serving the political purposes of government agencies and the economic purposes of ranchers and their creditors. On both dimensions, however, the Clementsian assumptions turned out to be flawed. The ecological variation within and between rangelands across the western United States, coupled with the practical and conceptual problems of studying those lands scientifically, confounded the needs and expectations of the state and capital alike.

If fault is to be found, it lies not with Clements or the range scientists but with the needs and expectations placed on them, which conflicted in fundamental ways with what the scientists were tasked with understanding—namely, the ecological processes at work in raising livestock on rangelands. Superficially, the processes look simple enough: plants grow on a piece of land; animals eat the plants, breed, and produce offspring. Grasses and other forage plants have evolved in the presence of grazing animals (and fire) for tens of millions of years, and they are adapted to withstand the periodic removal of portions of their aboveground parts; many species actually benefit, at least indirectly, from such disturbance (Sayre 2001). As long as there's enough forage, then, and water to drink, the whole system appears self-perpetuating. Hiding within this simple picture, however, is an extraordinary amount of complexity. Range scientists encountered this from very early on, and their impulse was always to collect more data. But at the heart of the matter was variation—the more something differs across space and time, the more intensive one's measurements need to be—and the more data they collected, the more variation they detected, so the problem grew more rather than less difficult. To begin to see why, it is useful to consider

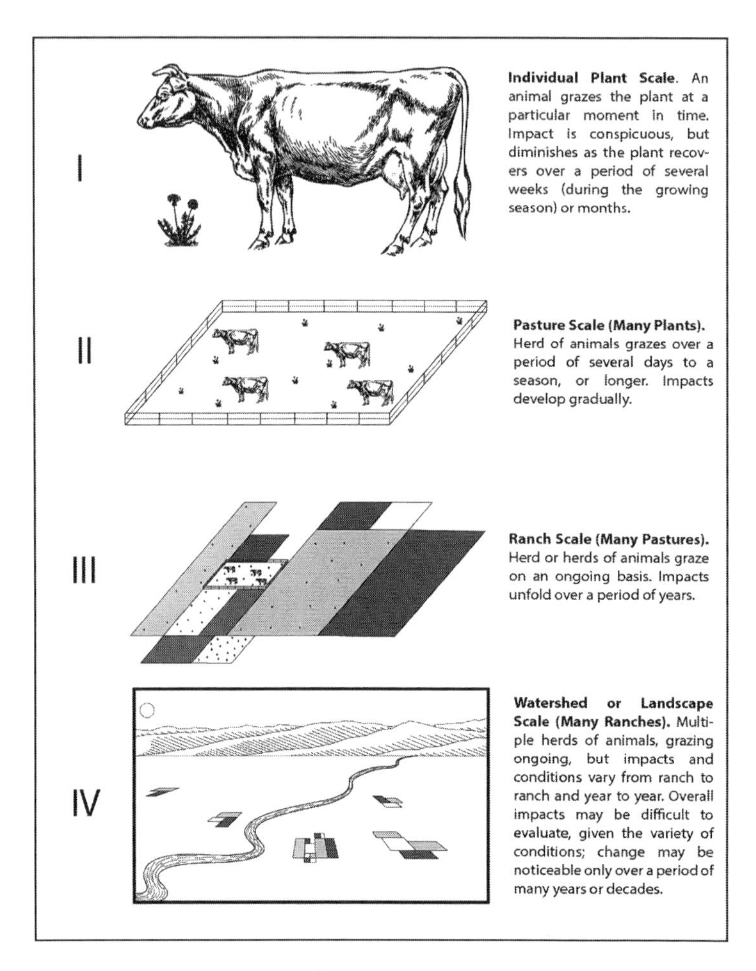

I.1. Four spatial scales for studying livestock grazing. Adapted from Sayre (2001, 13) by Xander Lenc.

the multitude of spatial and temporal scales involved. Figure I.1 depicts the four scales at which livestock grazing has most often been analyzed and understood: the plant, the pasture, the ranch, and the watershed or landscape. One could add more scales—for the molecular level of plant growth and animal nutrition, for example, or for entire regions or biomes—but these are sufficient for our purposes here.

Livestock not only need enough forage, but they need it every day, and their nutritional needs vary over time: females don't breed if they're poorly fed, for example, and they need more nutrition when they are nursing. Meeting these needs, then, depends on matching them to plant growth in time as well as space. Meanwhile, the nutritional value of grasses for livestock also changes over time: it is much higher when the plants are actively growing (and therefore green). Some plants, known as annuals, go through their entire life cycle—germination, growth, reproduction, and death—in a single year; others, known as perennials, live for multiple years (the life span of most perennial grasses is not more than about a decade). In many rangelands, annuals and perennials coexist, albeit in widely varying proportions. Unlike annuals, perennials retain fairly high levels of nutritive value when they are dormant (brown) during the non-growing season (Alvord's "philosophy of the cured grasses"), albeit much lower than found in green plants. Moreover, perennials are usually capable of surviving a dry year when annuals might not grow at all. Perennials are therefore generally considered more valuable plants on most rangelands. (This may also reflect the enduring influence of Clementsian theory, according to which perennials are a higher stage of succession than annuals.) But annuals typically grow more quickly and earlier in the growing season, making them valuable at a key point in time, when most perennials are still brown; they also tend to produce larger volumes of forage per unit of moisture. In warmer rangelands such as those of the southwestern United States, moreover, there may be two growing seasons, with annuals more active in winter-spring and perennials in the summer.

Just at the scale of the individual plant and animal, then, processes at several different temporal scales operate simultaneously: intra-annual for the supply and demand of nutrients; annual for some plants and the reproductive cycle of cows (gestation takes about nine months); and multiannual for other plants. Still more processes come into play, encompassing but exceeding those at the plant scale, when we expand to the scale of a pasture, ranch, or landscape. One must consider, for example, the qualities of soils and the availability of seeds, shaped by wind and water moving in patterns determined by topography, plant cover, and weather. If they are not densely concentrated, livestock graze selectively: the plants they choose to eat, how much, and when are not fixed, but depend on what other plants are available, how difficult it is to get to them (in terms of steepness of terrain and distance from water, for example), and how familiar the livestock are with the piece of land in question (which they learn from their mothers as well as from direct experience). The genetic makeup of a herd—controlled over multiple generations through careful selection of bulls, culling of cows, and retention of female calves for future breeding—also affects how the animals utilize rangelands. Climate cycles and the growth of longer-lived plants like mesquite, as well as sudden events such as severe floods or fires, may have effects that persist for decades or more. Finally, the timing and sequencing of these various factors—whether and how it rains following a fire, for example—can decisively affect the outcome.

As figure I.1 indicates, each spatial scale has a corresponding, most relevant temporal scale, and these align in a rough pattern: shorter durations for smaller scales, longer ones for larger scales. Historically, range scientists (like scientists in general) have tended to focus their research at a particular scale of space and time, choosing their units of analysis and conducting their studies accordingly. The majority of range science research historically was done at the pasture or ranch scales (II and III in figure I.1); research at the plant and landscape scales was conducted primarily by scientists in other fields. This division of labor obscured the fact that what happens at one scale is shaped by processes at other scales, and it is now understood that many of these relationships are not linear. Scaling up or down is not a simple quantitative matter, because different processes are determinative at different scales, and because a small incremental change in a variable at one scale may trigger qualitative changes at another. Indeed, the ecological effects of grazing vary depending on the scale at which one chooses to examine them (Fuhlendorf and Smeins 1999). Thus, the seemingly simple act of an animal grazing off a portion of a grass plant can have wildly different impacts—beneficial, benign, or destructive—depending on when it occurs, how much is grazed, how soon it happens again, and the larger historical and geographical context in which it takes place.

Two conceits tacitly informed range science from its inception. The first was that range livestock producers did not know how to manage their lands and herds properly. This seemed obvious, given the damage of the cattle boom and the recent arrival of many ranchers from other, less arid regions. But it ignored the potential of practical experience to yield valuable knowledge, as well as those people—many of them of non-Anglo heritage—who did in fact have such experience. The second conceit was that reductionist scientific methods could produce the knowledge that range livestock producers needed more quickly and reliably than could be achieved in other ways. This faith had two parts: first, that controlled, replicable experiments could isolate causal relationships—how much grazing a certain species of grass could tolerate, for example—by excluding or holding constant all the other relevant variables; second, that the resulting knowledge—that, say, 50 percent of a certain species of grass could be safely grazed—would reliably hold true at other times and places, without regard to history and geography. Part one was plausible to some degree, but part two—the abstraction of "scientific" knowledge from space and time—was a leap of faith enabled by faulty assumptions about scale.

The paradox of US ranching today is that it is the most ecologically sustainable segment of the country's beef industry, yet it is also the most economically marginal and historically the most vilified by environmentalists. Almost by definition, rangelands persist to the extent that people have not found ways to convert them to other kinds of land; they are used for livestock grazing by default as much as design. The decisive factor is whether the returns from conversion can justify the costs, which depends on the availability of capital, the relative returns of current and prospective uses, and the tools and techniques for converting the land. Science has played a leading role in developing those tools and techniques, largely with funds provided by government, and the aspiration to find economical methods of altering rangelands to enhance livestock production has been both abiding and dynamic. But the rangelands themselves have eluded many, if not all, of these attempts at human control, and in the United States, the most widespread practices—fencing, rodent and predator extermination, and artificial improvement—have been paid for largely with public funds. Rangelands have been changed in various ways—from erosion and invasive plants to the reduction or elimination of bison, beavers, wolves, prairie dogs, fire, and indigenous people—but it remains the case that they are only weakly and partially susceptible to intentional human manipulation. This reality has frustrated modern political and economic aspirations for more than a century. It is also what makes rangelands a continuing source of beauty and fascination as well as a lens through which larger lessons can be discerned.

ONE

Producing the Range

Extermination and Fences

In the fall of 1894, the USDA's Division of Economic Ornithology and Mammalogy sent J. Ellis McLellan to the southern San Joaquin Valley in California "to obtain all the information possible on the subject of rabbit driving" (Palmer 1897a, 53). The area had recently seen a large influx of settlers, drawn by the prospect of agriculture on lands previously used for livestock grazing, and the new residents had developed an elaborate community tradition to combat jackrabbits, whose numbers appeared to have exploded in response to the augmented food supply. Armed with clubs, hundreds and sometimes thousands of people would form a semicircular line around as much as five to ten miles of rangeland. On a signal, they would converge toward a V-shaped pair of low temporary fences, beating the bushes and scaring the jackrabbits into a small corral at the base of the V. If a rabbit turned back, it was clubbed to death; those that kept going met the same fate when the line of pursuers closed in on the corral (figure 1.1, top). One such drive, in Fresno County in March 1892, was reported to have resulted in the death of some 20,000 rabbits (figure 1.1, bottom). By 1897, McLellan and other government agents had documented 190 rabbit drives in the southern San Joaquin Valley over a ten-year period, with an estimated death toll of 470,600 rabbits (Palmer 1897a, 58).

Rabbit driving itself was not new. Native Americans had employed a similar technique to hunt rabbits for food at locations from Washington to Arizona, including in California. Farmers in the San Joaquin Valley had organized drives using shotguns in the 1870s and later switched to clubs after "several people were peppered with shot" as the hunters converged (Palmer 1897a, 53). A white hunter in Tulare County began driving rabbits in 1882 to capture them alive for sale to coursing clubs in Los Angeles and San Francisco; the clubs paid one dollar apiece for rabbits to release in staged

Bull. 8, Div. Biological Survey, U. S. Dept. Agriculture. PLATE III.

A JACK RABBIT DRIVE NEAR FRESNO, CALIFORNIA, MAY 5, 1894.—RABBITS ENTERING THE CORRAL.

Bull. 8, Div. Biological Survey, U. S. Dept. Agriculture

PLATE IV.

RESULT OF THE GRAND ARMY RABBIT DRIVE AT FRESNO, CALIFORNIA—20,000 JACK RABBITS KILLED.
(From photograph by Stiffler.)

1.1. Rabbit drives in California's San Joaquin Valley. Top: fencing and large numbers of volunteers were used to corral the animals, which were then clubbed to death. Source: Palmer 1897b. Bottom: The Grand Army Rabbit Drive in Fresno resulted in the deaths of some 20,000 jackrabbits. Source: Palmer 1897b.

hunts with horses and dogs. What *was* new was the scale of the drives, which reflected their goal: not simply to harvest some animals for food or sport but to exterminate as much of the rabbit population as possible. The famous California ranch of Miller and Lux reportedly paid its cowboys five cents for each rabbit killed in 1887, removing some 7,000 animals that year. Drives were also reported in recently settled parts of Oregon and Idaho, as were organized hunts with shotguns in Utah and Colorado. As unwelcome competitors for grass or crops, all rabbits were pests.

In his 1897 report, assistant chief of the Biological Survey T. S. Palmer opined that rabbit driving was probably on the wane in California since so many rabbits had already been killed and settlement had become so dense that their numbers seemed unlikely to rebound. He saw this as a good thing not for the sake of the rabbits but for the sake of the people's civility. "In the San Joaquin Valley a large drive is made the occasion of a general holiday; the schools are sometimes closed and women and children join the throng to assist in clubbing the rabbits or to watch the slaughter. It may well be questioned whether such scenes of butchery can have anything but an injurious effect on a community, and it will be fortunate when the necessity for them no longer exists" (Palmer 1897a, 59).

The rabbit drives did not push jackrabbits to extinction—the species persists in the San Joaquin Valley and throughout most of the West—and, in a broader sense, they were merely an extension of an impulse as old as domesticated agriculture itself: to target and destroy any organism that threatens one's crops or livestock. For all their brutality, the rabbit drives themselves are less noteworthy than the fact that beginning in this period, they attracted the attention of a federal government agency that sought to apprehend socio-ecological relationships in a certain way. Palmer's report included the basic natural history of the jackrabbit—its description, range, subspecies, breeding, and feeding habits—alongside an assessment of the "injury to crops and means of protection," poisoning, control measures in Australia, and bounty laws in nine western states. The perspective taken was at once scientific, economic, and didactic, with photographs, tables of data, and diagrams all presented with an air of clinical objectivity.

The jackrabbits of the San Joaquin Valley were not alone in being targeted. Exterminating economically damaging organisms was popular and widespread in the United States at the time, whether the target was "pests," such as rabbits and prairie dogs, or predators, such as wolves, bears, mountain lions, coyotes, and raptors. The number and variety of extermination campaigns—whether successful or merely attempted—is remarkable in hindsight, even though government agencies continue to persecute many such organisms to this day. Ordinary citizens, such as those in the Central Valley, initiated most of these campaigns, which extended and inverted the market-driven killing of commercially valuable wildlife that had already wiped out bison on the Great Plains and beavers from large swaths of the United States.

Nor was McLellan alone in producing this novel kind of knowledge. The closing decades of the nineteenth century saw a proliferation of federal government agencies charged with the study of the nation's natural resources, especially within the USDA. The Section of Economic Ornithology, established in 1885, studied how to control birds that damaged crops; mammalogy was added to its name when the section became a division in 1886. Ten years later it was renamed again, as the Division of Biological Survey; it became a bureau in 1905, and eventually came to encompass wildlife conservation when it turned into the US Fish and Wildlife Service, in the Department of the Interior, in 1934. Other branches focused on agricultural crops: the Divisions of Seeds (founded 1868), Botany (1869), Pomology (1886), and Vegetable Pathology (1890) were consolidated in 1900 into the Office of Plant Industry, which became a bureau in 1901 when it absorbed the Division of Agrostology—the unit that had been created in 1895 to conduct the first scientific research on rangelands and grazing. Many of these agencies engaged in extermination, first by cataloging and describing citizen-led campaigns and then by stepping in to make them their own, seeing opportunities both to improve the methods employed and to satisfy and build constituencies, especially in the recently settled and still-developing West. The study and management of trees and timber went through a similar series of bureaucratic transformations, beginning with a special agent in the USDA in 1876, then a Division of Forestry in 1881, and culminating in 1905 with the creation of the Forest Service, which would take the eradication of fire as one of its core missions.

Range science was born from and within this context of burgeoning government-scientific endeavors, against a backdrop of systematic extermination. For rangelands during most of the twentieth century, the organisms to be exterminated fell into two main categories: pests—mainly rodents—which consumed forage and/or crops, and predators, which threatened livestock directly.[1] Government campaigns to exterminate both kinds of "noxious" animals developed in the early twentieth century within the Bureau of Biological Survey (BBS), and they were only loosely coordinated—if at all—with early studies of rangeland ecology. But both would play significant and underrecognized roles in the future of the field.

One of the core challenges the BBS faced in both rodent and predator control was the mismatch of scales between the targeted organisms and the spatial division of rangelands among public and private landowners. Rodents (as populations) and predators (as individuals) routinely crossed boundaries and jurisdictions, so effective extermination required coordinated action at scales larger than individual parcels, states, or types of public lands. Not only would prairie dogs and predators have to be hunted, trapped, or poisoned, but people would also have to be persuaded or coerced into permitting BBS personnel to conduct extermination on their lands. In the name of making rangelands produce livestock more efficiently, organisms that had occupied those lands for thousands of years were deemed not only unwelcome but unnatural and incompatible with the ideal climax conditions of range vegetation that were subsequently enshrined in range science.

This chapter looks first at campaigns to exterminate rodents, particularly prairie dogs. These efforts extended throughout the western United States, and they had lasting effects on rangelands by removing prairie dogs from nearly 100 million acres where they had previously played important ecological roles (Oldemeyer, Biggins, and Miller 1993). I focus on extermination campaigns in Arizona and New Mexico, both to take advantage of the excellent unpublished dissertation by Claudia Oakes (2000) and because this area overlaps with the focus of subsequent chapters. I then turn to predator control and the most influential early experiment in the history of range science. Conducted between 1907 and 1909, the Coyote-Proof Pasture Experiment was a joint effort between the Forest Service and the Bureau of Plant Industry (BPI), which had hitherto been in charge of range research for the USDA. The name was somewhat misleading, because the goal was to exclude *all* predatory mammals, not just coyotes. Although it was not the first scientific experiment in range management, as is sometimes claimed,[2] it was the first to be deemed successful, and its results helped transform the very institutions that had produced it. Inspired by this perceived success, the Forest Service soon took over range research from the BPI, and the young scientist who conducted the experiment, James T. Jardine, became inspector of grazing, while his collaborator, Arthur W. Sampson, went on to become "the father of range science." (I will examine Sampson in detail in chapter 2.) The Coyote-Proof Pasture Experiment thus had enormous implications for how rangelands would be studied and managed for the rest of the twentieth century; indeed, few scientific experiments in history have influenced more land.

Here again, scale was a central challenge, but in different ways. How could the technique of fencing pastures, borrowed from moister areas in the eastern United States and Europe and developed at scales of one to forty acres, be made to work at the scale of thousands or even hundreds of thousands of acres in the more arid West? And how could the results of a single experiment, conducted on four square miles in the Wallowa Mountains of eastern Oregon, be extrapolated to other national forests, let alone the enormous and diverse settings of the West as a whole? These questions were not answered, or even seriously asked at the time. But the experiment nonetheless catalyzed the institutionalization of a model of rangeland administration and management that presupposed the combination of fencing and predator control. The model spread across the public lands of the western United States in a matter of decades and was subsequently exported to the developing world. It rested on weak scientific foundations, but it spread for other reasons, enabled by large public subsidies over many decades, especially in the form of labor under Depression-era jobs programs. This was ironic, because reducing labor—in the form of herders—was the semivisible, ulterior motive behind the experiment in the first place.

The aggregate result of extermination and fencing was a sociospatial order that turned western rangelands into the "range," as it would subsequently be understood in range science. As such, it produced a number of persistent and significant blind spots in the emerging field. The extermination of rodents and predators—as well as the removal of fire, which I will examine in the next chapter—altered rangelands in numerous, significant ways. But early range scientists tended not to recognize or acknowledge these changes, not only because what was absent went unremarked but because the bureaucratic division of labor between and within agencies helped obscure them from view: even if scientists in one agency were aware of the changes caused by another agency (which is unclear in most cases), they were not in a position to object or interfere. Most instead believed or assumed that their inquiries revealed the "natural" conditions and dynamics of rangeland ecosystems; they did not question the removal of pests and predators any more than they questioned the introduction of livestock. Fencing created a blind spot of another kind: range science would henceforth take for granted the division of rangelands into discrete, bounded areas, each with a determinable number of livestock belonging to a single owner. Herders, who were common across the West in this period, would give way to fences to reduce labor costs in a part of the country where wages were relatively high, and common or collective use of land would give way to exclusive use guaranteed by private ownership or lease. In other words, range science would be a science of ranching, not pastoralism (Ingold 1980). These blind spots would become fully evident only decades later (chapter 7).

Pests: Exterminating Prairie Dogs

Rodent "control" embraced everything from shooting, trapping, and flooding to a wide variety of poisoning techniques, and its targets included rabbits, squirrels, gophers, hares, rats, mice, muskrats, moles, woodchucks, hedgehogs, beavers, and porcupines (Cameron 1929, 55; Young 1936). But it was the prairie dog that prompted the most widespread and systematic extermination campaign of all. Of the five species of prairie dog (*Cynomus* spp.) identified in western North America, the black-tailed prairie dog (*C. ludovicianus*) was by far the most widespread, with a range that originally extended from northern Mexico to southern Canada. Perhaps as many as 1 billion in number, they inhabited colonies or "dog towns" that could extend for miles. Early explorers and settlers often found the animals charming and cute, although shooting them was also something of a pastime (Oakes 2000, 78–79).

Ecologists today consider prairie dogs a "keystone species" because of the multiple roles they play in semiarid grassland ecosystems: their burrowing activities cycle nutrients throughout the soil profile; the burrows themselves aid in water infiltration and provide habitat for a wide range of birds, reptiles, and other mammals; and, above all, prairie dogs are food for predatory animals such as hawks, owls, eagles, coyotes, foxes, badgers, snakes, and ferrets (Oldemeyer, Biggins, and Miller 1993; Sierra-Corona et al. 2015). With so many animals trying to eat them, prairie dogs evolved collective and social behaviors to protect themselves, including a set of vocalizations to warn one another of various kinds of threats. But it was another evolved behavior that earned prairie dogs the wrath of European American settlers and the BBS: the practice of eating all the vegetation in and around their colonies down to bare ground so they could see approaching enemies better. For newly arrived ranchers, this was wasted grass that should instead be consumed by livestock.

By the turn of the century, biologists and settlers increasingly classified prairie dogs as destructive pests (Oakes 2000, 84), and the fledgling Division of Biological Survey had already invested significant effort in perfecting prairie dog extermination techniques and educating landowners on how to use them. In one of the earliest government reports on the subject,[3] the famous naturalist and chief of the survey, C. Hart Merriam, noted strong popular interest and support "from the Dakotas to Texas," with some ranchmen willing "to pay for the destruction of the animals at a rate per acre exceeding the actual market value of the land" (Merriam 1902, 265). Prairie dogs not only consumed valuable grass and crops but threatened levees and irrigation ditches by digging burrows in them. Merriam attributed the problem to the very success of settlement: "The white man cultivates the soil and thus enables it to support a larger number of animals than formerly; at the same time he wages warfare against the coyotes, badgers, hawks, owls, snakes, and other predatory animals which had previously held the prairie dogs in check" (Merriam 1902, 263). He calculated that "32 prairie dogs consume as much grass as 1 sheep, and 256 prairie dogs as much as 1 cow," and from various accounts he estimated an average density of twenty-five prairie dogs per acre in dog towns. Describing a famous colony in Texas that covered roughly 25,000 square miles, he deduced that "the grass annually eaten by these pests in the great Texas colony would support 1,562,500 head of cattle. Hence, it is no wonder that the annual loss from prairie dogs is said to range from 50 to 75 per cent of the producing capacity of the land and to aggregate millions of dollars" (Merriam 1902, 258). Calculations like this one—deploying ratios and equivalences to translate plants and animals into revenues or opportunity costs relative to land values—were characteristic of the kind of official knowledge we will encounter throughout this history. They were routinely used, for example, to justify the Biological Survey's budgets: the 1924 annual report estimated that "noxious creatures" caused some $500 million of crop damage nationwide every year (Cameron 1929, 55).[4]

Merriam's report reflected a tension he struggled with throughout his time at the survey, between basic or "scientific" research to advance knowledge and applied or "economic" work to address practical problems. Merriam preferred the former, and from the beginning of his employment at the USDA in 1886 (when he became chief of the Division of Economic Ornithology and Mammalogy), he prioritized collecting specimens, describing species, and mapping their geographical ranges throughout North America. In this vein, the first half of his prairie dog report focused on natural history, describing the animal's range, habits, burrowing behaviors, and natural enemies, and he counseled in passing that some natural predators, such as hawks and owls, should be protected to help control prairie dog populations. But by 1902 he was under growing pressure from Congress and USDA administrators to demonstrate the practical benefits of the survey (Cameron 1929, 21–36), and the second half of his report provided detailed instructions for wholesale prairie dog extermination.[5]

Destroying isolated prairie dogs and their colonies was not difficult. Various methods had been devised and proved successful on small scales—including "trapping, drowning, destruction by domesticated ferrets, and capture in sand barrels and straw barrels placed over holes" (Merriam 1902, 266). But the more relevant scale was much larger—"areas comprising thousands of acres"—and costs could quickly spiral out of control. "The cost on large ranches should not exceed 18 cents per acre, and should fall as far short of this as possible" (Merriam 1902, 265–66). Only poisoning and fumigation showed promise at large scales. Strychnine and cyanide of potassium, mixed into grain baits and scattered across colonies at one tablespoonful per burrow, cost only 14.5 cents per acre for a first application, which "if carefully made in late winter or early spring when food is scarce, may be counted on to kill 75 to 80 per cent of the animals." The surviving prairie dogs could then be wiped out with a second, much cheaper application or "killed by bisulphide of carbon"—that is, fumigation of burrows with poisonous gas (Merriam 1902, 267).

Unlike predatory mammals, individual prairie dogs are relatively sedentary, often spending their entire lives within an area of one square mile or less. But in aggregate, even against the best poisoning techniques, they posed a similar problem of scale relative to the land tenure patterns of the West. Exterminating them on one parcel was futile if they persisted next door, since they would quickly recolonize the cleansed property. "The chief obstacle to the extermination of prairie dogs on the plains is lack of cooperation among landowners. . . . Many ranchmen who have again and again poisoned those on their own lands have finally given up in despair because of the rapid overflow from adjoining lands, new animals continually taking the places of those killed, until the expense and labor of repeated poisonings were too great to be continued" (Merriam 1902, 269).

The need to coordinate extermination across multiple properties, states, and types of landownership soon emerged as the principal rationale for empowering the BBS to lead the efforts, even though wildlife that did not cross state lines were legally under the jurisdiction of the states. Bounties had been the primary tool for controlling wildlife at state and local levels for decades, but the high costs and perverse outcomes that bounty systems could incur had begun to backfire: it was said, for example, that under an Iowa bounty, farmers could make more money raising coyotes and killing them than by raising sheep (Palmer 1897b, 64). When World War I made food production into a national security issue, wild animals were cast as a mortal threat, whether they consumed crops, forage, or livestock. Federal legislation passed in 1914 mandated that the BBS lead extermination campaigns rather than merely advise and instruct the efforts of others; also, the term "noxious animals," used in previous statutes, was changed to "wolves, prairie dogs, and other animals injurious to agriculture and animal husbandry" (Cameron 1929, 59), suggesting the prominence of these two animals in motivating the new program.[6] The following spring, $125,000 was appropriated to pay for predator control in national forests and the public domain, in addition to $280,000 for controlling rodents and predators on lands of all kinds. An outbreak of rabies across four states in 1916 further inflamed sentiment against wild animals (coyotes in particular), prompting Congress to appropriate another $75,000 for the survey's work (Cameron 1929, 47). The BBS ramped up its predator control program and expanded rodent poisoning in western states from 6.2 million acres in 1916 to 19.2 million acres in 1920 (Bell 1921a, 432).

Summarizing prairie dog control efforts as of 1910 in Kansas, where 2.5 million acres had been treated with poison over the preceding decade, Sheffer (1911, 118) declared that "the day of the dog is in its closing hours." But success would not prove nearly so easy. With its augmented budgets, the BBS set up large-scale demonstration projects on federal lands throughout the West. One such project took place on the Jornada Range Reserve in southern New Mexico from 1916 to 1918, shortly after the reserve had been transferred from the BPI to the Forest Service (see chapter 3). James Jardine wrote to the Jornada's grazing assistant, C. L. Forsling, in 1917, that "we are anxious to have both rabbits and the prairie dogs exterminated in order to make the Jornada Reserve a demonstration of range development and range management and improvement along all lines" (Oakes 2000, 158). But the experiment was considered a failure, even though prairie dogs were extirpated from the Jornada (and have never returned): "It took 3 years and a large labor force to poison less than 15,000 acres. In the end, the campaign relied on burrow-by-burrow application of expensive carbon bisulfide gas, a method that was not considered an alternative for large-scale applications" (Oakes 2000, 158–59). The Jornada extermination program did show the importance of making accurate pretreatment maps of prairie dog colonies, which soon became standard BBS procedure throughout the West. But it was not integrated into ongoing range science research at the Jornada—no mention of it can be found in BPI or Forest Service reports from the time or in two otherwise comprehensive historical summaries (Ares 1974; Conley and Conley 1984)—and the effects of prairie dog extermination on vegetation were never analyzed (Oakes 2000, 159).

The sheer extent of the species' range meant that scores of millions of acres would have to be treated, and unless extermination was complete, colonies could rebound in a matter of years. At the Jornada and elsewhere, prairie dogs displayed an unanticipated capacity to learn—refusing to take the bait after observing their brethren perish—so it was found necessary to precondition them by dispersing clean grain before returning with the poisoned variety, significantly increasing the costs of control (Oakes 2000, 161). More than 15 million acres were poisoned in Arizona and New Mexico from 1914 to 1933, at a cost of over $2 million (Oakes 2000, 179–81), and the BBS's Arizona office proudly announced that the last prairie dog in Cochise and Graham counties had died on June 25, 1922. But this proved premature, forcing the state office into bureaucratic sleight of hand: kangaroo rats—which were found throughout the state—were quietly added to the list of targets to justify continued poisoning campaigns, and reporting procedures were modified: "Instead of listing the acres treated for each rodent species, the Arizona BBS began reporting only the combined acreage of all rodents. These steps allowed the Bureau to continue to perpetrate the myth of complete extermination of the Arizona black-tailed prairie dog in the agency's Washington, DC offices" (Oakes 2000, 177). In New Mexico, where prairie dogs were much more widespread than in Arizona, nearly 16.5 million acres were poisoned between 1914 and 1942—an area larger than was originally mapped as occupied habitat. One-third of this area was treated under New Deal programs, primarily by the Civilian Conservation Corps (CCC).[7]

Outside of federally owned lands, the BBS relied on cooperation from private landowners and state and county governments to implement rodent control projects. Many states appropriated funds to match BBS expenditures, and private landowners contributed their own funds or labor as well. A 1936 BBS report boasted that the federal government's share of rodent control costs was only about 25 percent (Young 1936, 2). Many ranchers and cattlemen's associations were enthusiastic supporters, to the point that federal officials worried internally about the appearance of undue public expenditure to benefit private landowners. But there was also widespread resistance to prairie dog extermination. Some Native American and Hispanic residents hunted prairie dogs for food, and a significant minority of rural landowners, including European Americans, simply didn't share the view that prairie dogs were "noxious pests" that should be destroyed—at least not at their expense. Many distrusted the federal government and refused access to their private lands on principle.

Arguing that extermination had to be total if it were to succeed, the BBS lobbied state and local lawmakers to make *not* poisoning prairie dogs illegal, "coercing uncooperative or absentee landowners to control prairie dog populations" (Oakes 2000, 182). BBS officials publicly characterized the rodents as allies of the Germans, determined to destroy the nation's food supply. The New Mexico Rodent Law of 1919 authorized the use of state and federal funds to poison prairie dogs throughout the state and assessed taxes on landowners to reimburse their share of the costs if they would not pay willingly. Similar legislation was passed in Arizona. But resistance persisted, especially among poorer residents, for whom the cost of the poisoned grain—even if subsidized by the government—was prohibitive. "Indian reservations were often forced into extermination programs, regardless of tribal willingness, and Native Americans were compelled to supply the labor force for application of poisoned grain" (Oakes 2000, 184). One BBS employee recalled Navajo women who ventured onto colonies to sweep up the poisoned grain and throw it away. The tax assessment scheme was legally convoluted and difficult to implement, and in many areas the sheriff had to be brought in to compel payment on the spot. The extent and effectiveness of southwestern prairie dog extermination programs was highly uneven, depending in part on the tenacity of local opposition and noncooperation (Oakes 2000, 191).

Ecological studies of prairie dogs and grasslands lagged well behind extermination, and within the USDA they provided a kind of retroactive rationale for continued poisoning. From 1918 to 1922, Walter Taylor of the BBS collaborated with the Forest Service and Frederic Clements to design and conduct an experiment at three sites in northern Arizona that were occupied by Zuni prairie dogs (*C. gunnisoni zuniensis*). With Clements's assistant, J. V. G. Loftfield, Taylor measured aboveground grass production in meter-square quadrat plots in four kinds of sites: grazed by cattle, by prairie dogs, by both cattle and prairie dogs, and by neither cattle nor prairie dogs. Prairie dogs were poisoned to remove them from "protected" plots, and in one case were reintroduced when they disappeared from an area where they were needed for the study. In a USDA bulletin, Taylor and Loftfield (1924, 11) reported that over the four years of the study, prairie dogs destroyed 69 to 99 percent of the annual crop of the three major forage grasses and 80 percent of the forage crop as a whole. Prairie dogs preferred the same grasses as cattle, and they could completely denude large areas of forage during drought periods. The bulletin construed such impacts as unnatural, however. Livestock grazing and predator control had "completely upset [the] original balance" between prairie dogs and grasses, allowing the rodents to proliferate. It followed that extermination campaigns were not disruptive of grassland ecosystems but were rather the opposite: "As an offset for these two modes of interference with the natural equilibrium, the Biological Survey and various cooperating agencies have undertaken systematic campaigns for the extirpation of the rodents. If utter destruction of the range grasses over great areas is to be prevented, these campaigns must be increased in scope and number" (Taylor and Loftfield 1924, 12–13).

Outside of official USDA publications, however, Taylor subsequently reached quite different conclusions. In 1930 he published "Methods of

Determining Rodent Pressure on the Range" in the scholarly journal *Ecology*, treating rabbits, wood rats, and three species of mice as well as prairie dogs. Taylor described the methods of measuring rodent populations—including "the placing of a line of spots of poisoned grain, and counting the resulting kill" (Taylor 1930, 526)—and experimental designs for measuring rodents' effects on grass production, including those used in his earlier study with Loftfield. As before, he concluded that prairie dogs and other rodents could consume the majority of forage on a site and that "rodent grazing is a strong factor in modifying range forage" (535). But he also highlighted the possible benefits of rodents to soil processes, for controlling insects, and as prey for other wildlife, although these benefits were "usually indirect and hard to appreciate" (540). "Rodents are an important part of the biotic communities which make up our grazing ranges," he concluded. "Often, unfortunately, rodents do not fit in to the requirements of good range management. . . . While the case for rodent control is in many instances conclusively demonstrated, there are others in which it is not so clear, and still others where control is undesirable" (541). But Taylor's reticence about rodent control and his call for more research did nothing to curtail the BBS's extermination campaigns, which were redoubled from 1933 to 1942 with cheap labor from the Civilian Conservation Corps and continued again after the war. By 1960, prairie dogs had disappeared from approximately 98 percent of their former US range (Oldemeyer, Biggins, and Miller 1993, 1).

With prairie dogs virtually exterminated throughout the Southwest, most scientists scarcely thought to study them, and their broader ecological significance was overlooked or neglected for decades. Other rodents soon took their place as the putative causes of undesirable range conditions, however. Increasingly alarmed by the spread of mesquite and other shrubs at the expense of perennial grasses (chapter 3), range scientists found that kangaroo rats aided mesquite establishment by scarifying the seeds and caching them underground (Reynolds and Glendening 1949). More generally, rodents and rabbits were seen as competing with livestock for forage and exacerbating any negative effects of grazing. A study conducted on the Jornada Experimental Range, for example, found that jackrabbits consumed up to 99.4 percent of the perennial grass crop on degraded rangelands, preventing recovery and making artificial reseeding unfeasible without accompanying elimination of the rabbits (Parker 1938). Indeed, the system of range assessment that emerged by midcentury defined range conditions in such a way that prairie dog towns were "degraded" by definition (Koford 1958, 65), even if prairie dogs had been present for hundreds or thousands of years.

Their absence therefore became a necessary component of "proper" management, institutionalizing extermination within the dominant paradigm of scientific understanding of rangelands. That prairie dogs eat (and sometimes kill) mesquite saplings, and that rodent control may therefore have contributed to shrub encroachment in the Southwest, was effectively invisible to range scientists until the 1990s (Havstad, Huenneke, and Schlesinger 2006, 292; Gibbens et al. 1992). The disappearance of fire from these landscapes was probably still more important as a driver of shrub encroachment (Weltzin, Archer, and Heitschmidt 1997), but prairie dog extermination likewise produced a blind spot for the nascent field of rangeland ecology.

Predators: The Coyote-Proof Pasture Experiment

As with rodents, predator control was a BBS mandate, generally conducted without the involvement of range scientists. It was as extensive, systematic, and varied in its techniques as rodent extermination, with results that ranged from highly effective (wolves) to largely ineffective (coyotes). Prior to their widespread demise, however, predators played a key role in the origins of range science. On May 9, 1907, C. Hart Merriam sent a short memo to Gifford Pinchot. "Dear Mr. Pinchot: Your proposition to build a wolf and coyote proof fence on the Imnaha[8] National Forest in Oregon is of great interest to us, and the Biological Survey will gladly cooperate with the Forest Service in any way possible to secure satisfactory results." Three sentences later, Merriam—who in his career described more than 600 species of mammals—concluded with a blunt recommendation: "After the fence is completed, all wolves, coyotes, mountain lions and wild cats should of course be killed or driven out before the sheep are brought in." He made no mention of the purpose of the project, and his own agency's involvement was quite limited.[9] But Merriam took an interest because the project was a scientific experiment, for which the fence was an important (and expensive) apparatus.

After Congress transferred the forest reserves to the Forest Service in 1905, Pinchot moved quickly to consolidate and expand his new agency's scientific capacity and expertise. The scientific knowledge then available for rangelands, such as it was, resided in the BPI, and Pinchot turned to his fellow patrician, Frederick Vernon Coville, a BPI employee who was also chief botanist for the USDA and honorary curator of the National Herbarium. As Pinchot succinctly put it years later, "Until the Forest Service developed a body of experts of its own, Frederick V. Coville was the first and the earliest authority on the effect of grazing on the forest" (Maxon 1937, 280). The two men had become close friends during a three-week reconnaissance of forest reserves in central Arizona in 1900, proposed and hosted by Alfred F. Potter, a prominent local sheep owner and officer of the Arizona Wool Growers' Association. Shortly thereafter, Pinchot recruited Potter to come to Washington as assistant forester in charge of the Branch of Grazing, and Potter and Coville authored key portions of the Public Land Commission's 1905 report, which served as the blueprint for the grazing lease system subsequently implemented by the Forest Service (Richards et al. 1905). "Coville's studies and Potter's administration of the regulations laid the foundation for the Forest Service's grazing policies" (Rowley 1985, 39).

The son of a banker in upstate New York and a graduate of Cornell, Coville is most famous today for his botanical work with the Death Valley Expedition of 1891 and for his pathbreaking research on the blueberry, which helped make it a commercial crop in the northeastern United States. Although less well known, his role in early range research and administration must be considered among his most enduring accomplishments. His work began in June 1897, when the secretary of agriculture instructed Coville "to make an investigation of the alleged damage to forests by grazing of live stock, more especially the effects of sheep herding in the Cascade Forest Reserve of Oregon." The General Land Office had banned livestock grazing on all forest reserves in 1894, believing it necessary to protect forest regeneration and reduce fires. The move provoked resistance from sheep and cattle owners throughout the West, and at the end of June 1897, new policies were announced rescinding the ban for cattle nationwide but retaining it for sheep, except in the reserves of Oregon and Washington, where "the continuous moisture and abundant rainfall . . . make rapid renewal of the herbage and undergrowth possible" (quoted in Coville 1898, 11; cf. Rowley 1985, 32–35). Coville was tasked with investigating the wisdom of this exception.

Sheep grazing was much more controversial than cattle grazing at the time. In part this was because cattle did not require herders for protection against predators: if a herd of cattle wandered onto the (unfenced) forest reserves, there was no way to hold anyone accountable for intentional trespass, so cattle owners were less angry with the 1894 ban (Rakestraw 1958, 373). Antisheep—and antisheepherder—prejudice was also deep and widespread in Anglo-American conservation circles. John Muir famously described sheep as "hoofed locusts," and his vision for the forest reserves excluded both nonwhites and laborers of all kinds (DeLuca and Demo 2001). Muir had strongly influenced a National Academy of Sciences report that provoked outrage among livestock associations early in 1897 (Rowley 1985). Coville's assignment may have been a response to this controversy, and his subsequent report firmly rebutted two common accusations: first, that sheepherders were "aliens" or noncitizens who represented "a comparatively low class of humanity" (Coville 1898, 12); second, that they were responsible for starting forest fires. Apparently believing that both of these claims were settled by the facts on the ground, he focused his analysis elsewhere, detailing the herders' management techniques, their knowledge of the value of different plants for sheep growth, and the variation in their skills.

Drawing on both ecological and ethnographic observations, Coville's report articulated the empirical and analytical framework that would later inform the Coyote-Proof Pasture Experiment. Sheep grazing should be permitted on the forest reserves, Coville argued, but proper administration required a careful understanding of the inter-relations of fencing, herding, and predators. He began by distinguishing between "farm sheep" and "range sheep," pointing out that whereas the former were fenced, the latter were herded "on the great areas of unfenced public or Government land, popularly known as the open range, the outside range, or simply the range. Because this land is not fenced, and because unprotected sheep would be liable to destruction by wild animals, especially coyotes or prairie wolves, these range sheep are accompanied and cared for by a man who is called a sheep herder, or simply a herder. As a matter of economy, each herder is intrusted [*sic*] with as many sheep as he can properly manage, commonly two or three thousand" (Coville 1898, 9). Overgrazing in the Cascades had occurred, Coville wrote, but it was neither ubiquitous nor long-standing, and it had not yet altered the composition of vegetation in the reserve. He attributed the damage he observed less to the herbivory of the sheep than to their physical movements. "The principal bad effects of overgrazing are to be attributed rather to trampling than to actual close cropping" (Coville 1898, 27). This could be prevented, he believed, by securing sheep owners' access to the range resource. Open access for all made each herder rush to use the range ahead of the others, "to get all the grass possible without reference to the next year's crop, for he is never certain that he will be able to occupy the same range again. Where the competition is close the difficulty of insufficient forage is increased by the haste of the herder in forcing his sheep too rapidly over a grazing plot, the result being that they trample more feed than they eat. So year after year each band skins the range" (Coville 1898, 50). A system of permits to graze specified areas would relieve this competition and also enable the government to impose other terms, such as stocking rates and rules regarding fires. In the event, this would mean installing fences—turning the range *into* pastures, albeit very large ones—and thereby collapsing the distinction from which he began. "The effect of a moderate amount of grazing on the lands of the reserve is the same as the effect of the judicious removal of a grass crop from a fenced pasture by grazing or from a meadow by cutting; namely, that a forage crop is secured without material detriment to the land and the herbaceous vegetation it bears" (Coville 1898, 26).

In late March of 1907, Pinchot and Potter met with C. V. Piper, chief agrostologist in the BPI, to discuss "the range investigations which they would like to have undertaken by this Bureau." In a memo the following day, Piper reported to bureau chief B. T. Galloway that the Forest Service had reduced stocking on northwestern sheep ranges by about 25 percent, "by agreement with the stockmen. It is however the desire of the Forest Service to increase the carrying capacity of these summer ranges and consequently the allotment of stock to each district as rapidly as possible." Experiments were needed to learn whether degraded ranges could be reseeded and to determine "what system of range management both with cattle and with sheep will best permit the more valuable native grasses to re-seed themselves and thus increase the carrying capacity and maintain it at a maximum." Piper recommended that J. S. Cotton, who had studied range conditions in central Washington (Cotton 1904), be put in charge of the experiments, but Pinchot apparently overrode this suggestion in favor of Coville. An interbureau agreement was reached that the BPI would cover Coville's salary and expenses, while the Forest Service paid the rest of the costs.

In mid-May, Coville headed west by train to set up the experiments, stopping along the way to gather more information and recruit the necessary personnel. In Lincoln, Nebraska, he met with Charles Bessey and his protégé, Frederic Clements, who recommended one of their students, Arthur Sampson. Coville interviewed Sampson and hired him immediately. Sampson's training with two of the world's leading ecologists was critical for Coville because it meant that the other men he hired would not need such expertise; its importance for the future of range science was even greater, as we will see in chapter 2. Five days later, in Logan, Utah, Coville hired James Jardine, whom he described in a letter to Pinchot: "Mr. Jardine is twenty-five years old, a graduate of Utah agricultural college, and now an instructor in that institution. He was brought up on a ranch in southern Idaho and is familiar with the handling of cattle, horses, and sheep. He is well qualified both by his personal characteristics and his training to take part in the forest grazing investigations."

En route from Lincoln to Logan, Coville stopped in Laramie, Wyoming, to interview a prominent sheep owner by the name of Francis S. King. An immigrant from England, King and his brothers owned or leased some 120,000 acres of private land in Wyoming and were famous for their prize-winning purebred Merino, Rambouillet, and Corriedale rams and ewes (Bartlett 1918, 59). Coville's notes from the meeting[10] suggest that King's ideas and opinions played a significant role in refining the details of the imminent experiments in Oregon. King saw two advantages to protecting sheep within a predator-proof enclosure: reduced labor costs and improved animal performance. "Lambing within a wolf and cat proof fence would be highly advantageous, because three men could lamb 2500 ewes, one to keep the drop herd bunched and moving, one to move each lamb and ewe to shelter, one to look after the lambs for a day or two. Mr. King now has to use 5 men with each half band of 1500 lambing sheep. These men are kept with the half band about 10 days. Such temporary help is often not trustworthy. On a wolf and cat proof range two men could care for 8000 to 10000 sheep." King reported results from his own experience to support the claim that sheep would gain more weight and produce more wool if fences replaced herders. The key mechanism was a change in the animals' behavior. "When lambs bunch together as they do when a band beds down [on the open range], they run off a large amount of fat in racing and playing. A bunch of 37 thoroughbred lambs run in a band averaged about ten pounds lighter than the same number of lambs the following year from the same dams and sires run in a pasture on grass not quite so good. Difference of $3.50 in price of lambs. The ewes sheared 2 to 3 pounds more."

The core problem, according to King, was the herding system itself, with all that it entailed: unreliable laborers moving sheep in bands or bunches across open range in the presence of predators. Coville noted "from 5 to 15 percent loss from herding and wild animals," and he listed numerous reasons to be rid of herding, including "Saving in herding; Carrying capacity of range, probably double; Condition and value of stock and wool; Sheep fatter, healthier, wool worth more." These were the expectations that Coville carried into the experiment when he arrived in Oregon in late May.[11]

On June 3, Coville reported to Pinchot that he had selected a site with good forage, ready access by wagon, and "an abundance of coyotes, wild cats, and bear, with an occasional cougar and lynx." "The stockmen" in the area, he wrote, "are greatly interested in the experiment, the consensus of opinion being that under a pasture system the carrying capacity of the range will be doubled." He had found a "thoroughly efficient hunter" to hire for the project, and he recommended some modifications to the fence design based on his conversations in Wyoming (presumably with King). By mid-June, Jardine and Sampson had reported for duty, the fence was under construction, and Coville soon returned to Washington.

1.2. The fence enclosing the coyote-proof pasture, Wallowa National Forest, 1907. Photograph courtesy of the National Archives.

The fence—an imposing and elaborate combination of barbed wire, wire mesh, and stout posts (figure 1.2)—was completed too late for a full season's research, but a band of sheep was placed in the new pasture for one month, and observations were made both of their behavior and of the effectiveness of the fence at repelling predators. As for the latter, "the fence proved successful as a protection against coyotes, not successful as a protection against grizzly bears, doubtfully successful against black and brown bears, still problematic against cats, and not successful against badgers" (Jardine and Coville 1908, 23). The hunter and his hounds patrolled the perimeter each day, searching for signs of predators and, when possible, pursuing them (figure 1.3). But just as Merriam had feared, one coyote was trapped inside the completed fence, and two sheep were lost to predators—one to a bear and another to the coyote (Jardine and Coville 1908, 21). These predators interfered with efforts to observe "the action of sheep when they are allowed perfect freedom in an inclosure [*sic*] and protected from annoyance by animals" (Jardine and Coville 1908, 25), but Jardine nonetheless judged that "the test was very satisfactory. They [the sheep] retained, more or less, the natural instinct to mass, but from the first day the tendency to open, scattered grazing, with little or no trailing, increased" (Jardine and Coville 1908, 29).

Even before real data were available, Coville expressed his views in the introduction to Jardine's first report, published the following year. Both herders and herding represented obstacles to the efficient use of range resources on the national forests:

> Even with the best herders it is impossible to handle large bands of sheep with the same grazing efficiency as is secured in the fenced pastures of the eastern United States, and when one considers the large percentage of herders who are not skilled or who have a greater regard for their own comfort than

1.3. The hunter and his hound, hired to patrol the coyote-proof pasture against predators during the experiment. Photo courtesy of the National Archives.

> for the interests of the owner of the sheep or for the permanent welfare of the range, the aggregate waste can be regarded in no other light than as a matter of serious public concern. That one-third of the vegetation on the sheep ranges in the National Forests is destroyed by trampling is regarded as a conservative estimate. (Jardine and Coville 1908, 5–6)

Herding was wasteful not only in uneaten forage but in meat and wool not produced due to needless expenditure of the sheep's energy. Estimating that sheep owners had grossed $7,225,000 from national forest grasses in 1907, Coville deduced that another $3.5 million had been foregone due to "grass trampled and wasted" (Jardine and Coville 1908, 6). He specifically blamed herders who returned every night to a single base camp, which consequently became denuded of vegetation, rather than "bed[ding] his sheep upon the range wherever they finish the day's grazing, and sleep[ing] beside them" (Jardine and Coville 1908, 5). The latter practice, known as "bedding out," became a staple of Forest Service sheep management recommendations, one tightly associated with Jardine's name. Coville asserted that such "improved methods are rendered practicable" through secure access to national forest rangelands under grazing leases, although he can only have meant that *enforcement*, not bedding out itself, would then be practicable. In any case, while conceding that solid data had yet to be collected, Jardine's report concluded: "That the experimental system will materially increase the carrying capacity of the range is not to be doubted" (Jardine and Coville 1908, 32).

The following summer, after repairs had been made to the fence, a band of 2,209 sheep were turned into the coyote-proof pasture without a herder. The hunter resumed his daily patrols, and the behavior of the sheep was monitored all day, every day from June 21 to September 24. Only fifteen sheep died during the ninety-six days, none due to predation. The objective was to induce "open grazing," as Jardine termed it. "Whether sheep are in large or small bunches it is essential for the protection of the range that they be well scattered and graze quietly. Close-bunched grazing, massing, running, and trailing one after another should be prevented if possible, not only for the good of the range but for the good of the sheep. In this respect there was marked change during the season" (Jardine 1909, 23). Here, the predators were not the only culprits: herders themselves, and especially herders' dogs, were like predators in that they could provoke fright in sheep and cause them to bunch, mass, and run. The fence made herders and dogs unnecessary, and, over the course of the summer, "the sheep gradually became accustomed to free, unmolested grazing, and forgot the habits learned when herded" (Jardine 1909, 23). They formed smaller bunches, both while grazing and also to sleep at night, and they moved more lightly over the landscape. "The result was that little or no damage was done to the forage crop in this way. The entire crop was eaten and not wasted" (Jardine 1909, 23).

By weighing twenty lambs "of average size" at the beginning and end of the experiment and collecting similar data from bands herded on the unfenced range nearby, Jardine compared animal performance under the two systems. In eighty-eight days inside the fence, the pastured sheep gained an average of twenty pounds, whereas those from a band on the outside gained only fifteen pounds, on average, in ninety-six days. Overall losses among four outside herds ranged from 1.4 to 3 percent, compared with just 0.5 percent for the fenced herd. Finally, the fenced sheep required only 0.0156 acres per sheep per day. Three open-range herds, roughly estimated, required 64 to 123 percent more acreage, which Jardine attributed to the effects of trampling (and, in one case, "poor herders, who were of French descent, unable to speak English" [Jardine 1909, 32]). "It is safe to conclude that range grazed under the pasturage system will carry 50 per cent more sheep than when grazed under the herding system, where the band is driven to and from camp each day" (Jardine 1909, 32).

In the closing pages of his report for the 1908 season, Jardine took up the economic question: "Will the pasturage system pay?" Here he faced a difficulty, because the coyote-proof pasture had been very expensive to construct: $6,764.31, to be precise, including more than $2,000 in materials, $1,000 in transportation costs, and $1,000 to clear heavy timber from portions of the fence line (Jardine and Coville 1908, 18–19). This amounted to nearly $850 per mile (equivalent to about $22,000 per mile in 2015 dollars). Jardine chose not to use these figures, however, arguing that the location was exceptionally remote and heavily timbered. He estimated that under more typical conditions, "the cost on most grazing lands will approach very closely $400 a mile" (Jardine 1909, 39). He then tabulated the financial benefits in increased carrying capacity, heavier sheep, reduced losses to predation, and lower labor costs. Not counting increases in the lamb crop and in the amount and quality of wool (which he considered to be certain but not yet measureable), Jardine arrived at an annual return of $746.50, based on a herd of 2,200 sheep in a 2,560-acre pasture for three months. Thus, an initial investment of $3,200 (for eight miles of fence), at 8 percent interest and including maintenance, would pay for itself and begin yielding dividends after six years. He omitted the costs of the hunter from his analysis.

The experiment was repeated in 1909 with 2,040 sheep enclosed in the pasture for ninety-nine days, during which time only four perished. The results were nearly identical to those from 1908, although Jardine's report was more detailed and emphatic in its declarations of the virtues of "the pasturage system." The hunter killed one grizzly, two badgers, seven brown bears, and seven coyotes (the grizzly was killed even though it made "no attempt to go through [the] fence"). Just one brown bear and three badgers managed to breach the fence (Jardine 1910, 8). The sheep again displayed a gradual tendency "to depart from their old habits and accommodate themselves to the freedom of the pasture," so much so that by the end of the season "it was almost impossible to keep them close bunched without a dog" (Jardine 1910, 18, 19). Losses to poisonous plants as well as predators were higher among sheep herded outside the enclosure, and the pastured sheep were again heavier at the end of the season than the herded sheep (although this may have been due to differences in breeding) (Jardine 1910, 25). Acreage required per sheep per day was 52 to 90 percent higher outside the fence, although Jardine attributed some of this to poor-quality herders (including the one of French descent again, though by this time he had learned a little—"very little"—English) (Jardine 1910, 27–28).

The issue of herder skill presented a puzzle, which Jardine acknowledged but declined to address directly. "A first-class herder will work all the time" and seldom use a dog, resulting in "quiet, scattered grazing that may approach the pasturage system in efficiency," whereas "a lazy man . . . will wear out his dogs, worry the sheep, and destroy the forage" (Jardine 1910, 31–32). With respect to weight gain, "there is as much difference in the results obtained by a first-class herder and those obtained by a poor herder as there is between the results under the pasturage system and those secured by the good herder" (Jardine 1910, 26). And while "range grazed under the pasturage system will carry from 25 to 50 per cent more sheep than when grazed under the herding system," it was also possible "that an excellent herder can, to a considerable extent, allow his sheep freedom and keep them quiet, thereby increasing the carrying capacity of his range. No doubt there are herders who do this" (Jardine 1910, 28). All told, "the carrying capacity of the same range utilized by different herders may vary at least 25 per cent" (Jardine 1910, 31).

These were potentially troublesome admissions to make, for both scientific and economic reasons. Herder skill was clearly an important variable in sheep performance, but it was one that Jardine could neither measure nor control. Removing herders might thus be seen as necessary to a properly "scientific" assessment of range grazing, and it could clearly affect any calculation of the economic rationality of building fences and controlling predators. What if better training for herders were a more economical solution? Instead of confronting these issues, Jardine reverted to general claims about labor needs under the pasturage system, superseding by half the estimate that King had given to Coville two years earlier: "It is probable that one energetic man . . . can properly care for four inclosures [*sic*] similar to the experimental coyote-proof pasture," meaning "one man would care for from 8,000 to 10,000 head of sheep" (Jardine 1910, 22). Notwithstanding these problems, Jardine reached the same conclusions in 1909 as he had the year before, even extending them to include lambing as well as grazing sheep (Jardine 1910, 40).[12]

The Wallowa experiment was immediately hailed as a remarkable success, and the Forest Service quickly embraced it as guidance for the administration and management of rangelands elsewhere. The "pasturage system" appeared to solve numerous problems and satisfy everyone, provided one ignored or excluded the herders. In his "Annual Report to the Forester for the Fiscal Year 1909–1910" for the Branch of Grazing, Alfred Potter pronounced the results "very gratifying" and summarized them as follows: "The primary objects of the experiment have been accomplished, i.e., it has been demonstrated that the grazing capacity of the Forest lands can be largely increased by improved methods of handling stock, and that the increased cost of such methods, if any, is offset by increases in the number and weight of lambs raised, heavier wool crops, and reduced losses from predatory animals." Notably, Potter omitted any reference to labor costs in his summation. He suggested that the results be applied "to spring and fall or yearlong ranges" in other national forests. Potter also tabulated the accomplishments of Forest Service personnel assigned to predator control: 269 bears, 129 wolves, 148 wolf pups, 1,155 wildcats, and more than 7,000 coyotes killed in the eleven western states, "an increase of 109 per cent over the number of animals destroyed last year" and representing "a total saving to stockmen of considerably more than one million dollars per year."

The effects of the Coyote-Proof Pasture Experiment extend to the present day and across literally hundreds of millions of acres of rangelands in the United States and elsewhere. But by today's standards, the experiment was far from impressive. Its findings were presented as scientifically robust ecological facts about measurable interactions among predators, livestock, fences, and vegetation, but the underlying methods varied in several ways from one year to the next, and they relied on qualitative assessments or herder accounts for some important data. Sheep breeds, herder practices, vegetation types, and other variables confounded the findings, and researchers made no direct attempt to assess the impacts of the pasturage system on vegetation. There were no real controls by which to judge the relative effects of the enclosed pasture, the absence of predators, and the absence of herders or dogs as factors influencing the dependent variables (weight gain and wool clip). Estimating the carrying capacity outside the fence was imprecise and effectively impossible, since those herds were not confined within fixed boundaries and their circuits were likely noncontinuous and overlapping. Moreover, the ultimate metric for evaluating success was not ecological but economic: costs and returns, and thus profit on investment, determined whether fencing and predator control were worth implementing. In this calculus, the decisive factor was neither fences nor livestock performance, but rather the labor of herders. The high cost of fencing could be justified economically only if the fences greatly reduced the need for herders—and in the absence of herders to protect livestock, predators would have to be rendered effectively insignificant. Here again, the methods were flawed, with key costs either minimized (the fence) or excluded (the hunter) to reach the desired conclusions. Finally, Jardine's reports were not subject to peer review, rather only to the scrutiny of his superiors in the USDA, who themselves appear to have prejudged the results. Virtually every finding was suspiciously similar to the expectations that Coville carried into the experiment in 1907. There is almost no chance that the experiment would be recognized as publishable, or even scientific, if it were conducted today.

The experiment succeeded not on the basis of its scientific rigor, however, but because it lent authority to ideas that were already viewed favorably within the institutional context that gave rise to it. Coville and Jardine produced a set of knowledge claims that appeared to conform to scientific norms of experimentation: the deliberate manipulation of objects, organisms, and people and the careful recording and interpretation of actions and reactions among them. Most of these primary data were of a broadly ecological character and thus appeared apolitical and objective. The results were translated into economic terms to assess the practicability of implementing a similar management regime on western rangelands as a whole, and if the economic analysis was at once flawed and decisive, this contradiction would eventually be resolved by government largesse: the cost of both fencing and predator control would be subsidized by an array of federal agencies over the decades to come.

Predator control occurred throughout the national forests and beyond, organized and funded by the BBS under federal legislation passed in 1914. Rather than exclude predators with expensive fences, the strategy became simply to exterminate them by hunting, trapping, and poisoning. The bureau divided the western states into eight "predatory-animal districts, each in charge of a predatory-animal inspector. The hunters employed devoted their entire time to the work, and were not permitted to receive bounties from any source. The skins of all animals having fur value taken by the hunters became the property of the Government and were sent in to the Department and sold at public auction, the receipts being turned into the United States Treasury" (Bell 1921b, 292). Between 1915 and 1920, the survey and its cooperators killed 128,513 predatory animals in sixteen western states by hunting or trapping, including 109,346 coyotes and 2,936 wolves; the number killed by poison was unknown but believed to be equal or greater (Bell 1921b, 299). The 1920 USDA *Yearbook* boasted: "Reports by stockmen indicate that on many ranges and lambing grounds the former heavy annual losses [to predation] have become negligible or have been entirely eliminated" (Bell 1921b, 298). Coyotes, black bears, and mountain lions persist throughout much of the West today, and wolves and grizzly bears are still found in parts of the northern Rockies, but their numbers are too small to pose a significant threat to livestock: less than 0.25 percent of US cattle, for example, are lost to predators, including dogs.[13] Many predators continue to be persecuted by the BBS's descendent agency, Wildlife Services; even those that are not persist only at much-reduced numbers. The magnitude of the influence of predators on ecosystem processes is controversial (Marris 2014).

Given the ubiquity of extermination across the agencies responsible for natural resources in the United States in the early twentieth century, one is forced to the broader conclusion that conservation itself was born at least as much out of the impulses and effects of exterminating "bad" things as out of the ideal of conserving "good" ones. As one scholar noted in 1929 in a semiofficial history of the Bureau of Biological Survey, "Broadly speaking the entire history of the Survey from the time it became a bureau down to the present day, is a history of the growth . . . along two main lines: First, the repression of undesirable and injurious wild life; second, the protection and encouragement of wild life in its desirable and beneficial forms. There are other lines to the story, but these are the principal ones, to which all the others are, in one way or another, subordinately connected" (Cameron 1929, 40). In some cases, extermination triggered epiphanies of loss and regret, as in Aldo Leopold's famous (and perhaps apocryphal) "green fire" in the eyes of a dying wolf (Leopold 1970, 137–41). During the Depression, extermination provoked a large group of prominent academic naturalists to form an Emergency Conservation Committee to protest "the vast machine for wild life destruction into which the misnamed and perverted 'Biological Survey' has gradually been developed since the war. For this machine new employment and new victims must constantly be found" (Edge and Emergency Conservation Committee 1934, 4). More often, though, extermination was simply a normal practice that went hand in hand with conservation, viewed as a practical necessity whose historical and philosophical significance was overlooked. Exterminating some organisms was the flip side of protecting others, requiring the same knowledge and expertise even if the tools for the two tasks differed; it persists to this day in efforts to combat invasive species.

Jardine predicted in 1919 that "fences eventually, no doubt, will be constructed to control the stock" on the national forests (Jardine and Anderson 1919, 21). He was correct, but it would require enormous public expense to realize. Fence construction was underwritten by the Forest Service;[14] the General Land Office's Grazing Service; and several New Deal programs, including the Agricultural Adjustment Act[15] and the Civilian Conservation Corps, whose crews built almost a million miles of fences nationwide between 1933 and 1942 (Maher 2008, 43).[16] As if to confirm the flaws in Jardine's original analysis, the fences were much simpler and cheaper, with just four strands of barbed wire instead of the elaborate Wallowa design. Building them to repel predators was far too expensive relative to the risks of predation once predators had been "controlled" by other means. The purpose of the fences, instead, was to make it possible to define and allocate public lands for private use by individual ranchers and, specifically, to control stocking rates: how many cattle or sheep were permitted to graze, for how long every year, under exclusive leases.

Not that such control was necessarily achieved on the ground any time soon: Forest Service staffing remained insufficient to enforce the rules across its thousands of large and remote grazing allotments, and in practice stocking rates could easily deviate from the official rules even after fences were finally built. For political, economic, and scientific purposes alike, however, actual practice was less important than the ideas and norms made possible by fencing. The exclusive use of land—as opposed to collective or common use—distinguishes ranching from pastoralism (Ingold 1980), and the spatial organization of rangeland livestock production into bounded, fixed areas, in which vegetation could be measured and monitored in relation to determinate amounts of grazing, became constitutive of what "range" would mean for most of the twentieth century.

The problem of scale was obscured by and within this sociospatial order. Although neither Coville nor Jardine said so explicitly, their vision of "open grazing" imagined that livestock impacts on rangelands could be made homogeneous, distributed evenly across space regardless of the size of a pasture (Fuhlendorf and Engle 2001). But at the scale of actual grazing allotments—most of them much larger than the coyote-proof pasture, let alone pastures in the East—heterogeneity in topography, soils, vegetation, and water supplies defied this aspiration. Giving sheep and cattle "freedom" from predators and herders allowed them not only to spread out but to select the places and plants they preferred, whether that meant concentrating along a shady creek, avoiding a steep slope, or eating only a few of the plants in a given spot. Most of these problems were compounded along the temporal axis, because livestock impacts could vary widely at seasonal, annual, and longer time scales, which the Coyote-Proof Pasture Experiment could not capture. As we will see in the next three chapters, these problems beleaguered range scientists and administrators in the United States for decades; the blind spots pertaining to fencing would not become fully evident until the ranching model was applied to pastoralist societies overseas in the second half of the twentieth century (chapter 6).

The ecological effects of fencing probably cannot be disentangled from other factors it has enabled or accompanied, such as water development, reduced herd mobility, and land tenure rationalization; fragmentation is now considered a major threat to rangelands worldwide (Galvin et al. 2008). In any case, fenced pastures and the near-total absence of large predators have by now been ubiquitous on US rangelands for so long that they are widely taken for granted not only by range scientists but also by the government and the general public. Fencing is effectively unquestioned as a basic tool of ranching and rangeland management, subsidized by US government agencies and aggressively promoted in pastoral development projects overseas; the western United States was the birthplace of barbed-wire fencing, which Netz (2004) considers fundamental to the "ecology of modernity." Today, "open range" no longer signifies the absence of fences altogether, but instead their absence along remote roadways, where livestock may imperil motorists without the livestock owner being liable for damages. Fencing and predator control are treated as separate issues in public debate and policy recommendations,[17] and the historical link between them is forgotten.

If both fencing and predator control were already planned or ongoing, and might well have proceeded without "scientific" support, then the most important legacy of the Coyote-Proof Pasture Experiment was its role in institutionalizing range research in an agency whose primary mission lay elsewhere. In 1910, the Forest Service created its own Office of Range and Forage Plant Investigations within the Branch of Grazing; five years later, all USDA grazing and range research was transferred from the Bureau of Plant Industry to the Forest Service. There is no indication that Coville objected to the transfer, and he probably could not have foreseen its longer-term consequences, but this seemingly minor bureaucratic shift had lasting and profound implications for the trajectory of rangeland administration and research. Jardine and Sampson were elevated from "special agents" to permanent positions in the Forest Service, from which they became the principal architects of the dominant paradigm affecting US rangelands in the twentieth century (Chapline et al. 1944, 131). Jardine's role was primarily administrative: as inspector of grazing in charge of the new office, he oversaw range research and directed the monumental task of "range reconnaissance" throughout the national forests over the following decade (chapter 4); he later served as chief of the Office of Experiment Stations and director of research for the USDA. His 1919 bulletin, *Range Management on the National Forests* (Jardine and Anderson 1919), was the first comprehensive statement of the policies and principles guiding forest rangeland management; it was still in use at his retirement in 1945—having been "three times reprinted without change"—when he was described as having "brought out the principles on which are founded the standards of good grazing practice over the whole western range country" (Rand 1945). Sampson, meanwhile, became "the father of range science," placing it firmly on the basis of Clementsian ecological theory, as we will see in the next chapter.

TWO

Fire and Climax

Bureaucratic Divisions of Scientific Labor

Like predators and rodents, fires were systematically suppressed by federal government agencies throughout the western United States beginning in the 1910s. Ecologists today agree that the negative consequences of long-term fire suppression are profound (Pyne, Andrews, and Laven 1996), but it took half a century or more for this knowledge to emerge. The Forest Service viewed all fires as contrary to all its objectives—timber production, watershed protection, wildlife conservation, and forage for livestock—and the primacy of timber over range was official policy for research as well as management. Moreover, as Pyne (1997, 191) notes, "The political agenda of the Forest Service dictated its scientific agenda. Administrative policy proposed the ends and research sought to supply the means." When confronted with research that suggested *any* benefits from fire, "the Forest Service systematically—shamefully—suppressed the data and sought to influence the publication of any results, even outside the agency, that contradicted official orthodoxy" (Pyne 1997, 197). Thus, the importance of fire in forest ecology remained obscure until scientists outside of the USDA finally managed to overcome the Forest Service's opposition to the publication of their results: Harold Weaver, Harold Biswell, and Harry Kallender, for example, all did their work with the Bureau of Indian Affairs or in universities; Biswell had to leave the Forest Service in order to pursue this line of research (Carle 2002). Even researchers elsewhere in the USDA were highly vulnerable. S. W. Greene of the Bureau of Animal Industry, for example, lost his job for publishing research in the 1920s and 1930s showing "that cattle gained more weight when grazed on burned range than on forest ranges protected from fire" (Carle 2002, 43).

It is well known that scientific forestry was imported to the United States from Europe—especially Germany in the late nineteenth century and France in the early twentieth—and that the European forestry model failed to recognize the importance of periodic fires for the functioning and persistence of North American forests (Pyne 1982). What has not been widely noted is that in most places, the frequency and size of forest fires were less a function of the trees than of the grasses that grew beneath and between them, providing the fine fuels in which recurrent fires could start and spread. European forestry's ignorance of—and prejudice against—fire, then, reflected its ignorance of grasses, which in the European context were deemed important only in "improved" pastures that were both spatially and intellectually segregated from forests. This division of scientific labor transferred to the United States as well: fire suppression proceeded without serious input from, or study by, range scientists.

With its incorporation into the Forest Service, range science became a sort of bureaucratic stepchild, both different from and subordinate to forestry, with profound and enduring effects. Had it remained in the Bureau of Plant Industry, the study of rangelands might have been able to proceed relatively uninhibited by other considerations, much as the Bureau of Biological Survey proceeded with predator and rodent control. Instead, bureaucratic divisions within the Forest Service—separating forestry from range, research from management, and fire suppression from the rest of the agency—constrained the kinds of questions that scientists could ask about rangelands and the kinds of answers they were expected to provide. Moreover, being marginal within the Forest Service played a decisive and poorly understood role in shaping the conceptual foundations of range science as a whole.

Fire

That livestock grazing could influence forest fires was recognized from the inception of the federal forest reserves in the 1890s, but mostly in connection with people rather than animals. Herders were routinely cast as "incendiarists" who burned the range on purpose to kill trees and shrubs and/or to encourage fresh green forage growth. In addition, many herders were of Hispanic or southern European heritage, and their critics deemed them unworthy, un-American, or even uncivilized in the racially charged debates of the period about public lands and the nation's forests. Herders typically ran sheep, moreover, and in most parts of the West, cattle owners were more white, wealthy, and politically powerful than sheep owners or herders. It seems certain that some herders did indeed start some fires, but there were plenty of other potential culprits, from lightning and hunters' ill-tended campfires to sparks flying from locomotives and railroad tracks. As we saw in chapter 1, Frederick Coville concluded that herders were not to blame for fires, based on his own detailed reconnaissance of the Cascade Mountains. But in the absence of solid evidence about the causes of particular fires, let alone data for fires nationwide, the debates were prone to hearsay and prejudice. Persuading people to prevent and suppress forest fires became a primary obsession of the Forest Service, symbolized by Smokey the Bear—one of the US government's most famous and successful propaganda campaigns, shot through with racial and nationalist sentiments (Kosek 2006).

Determining the effects of *livestock* on fire patterns was an even more complicated challenge. Filibert Roth, chief of the Forestry Division of the General Land Office, echoed Coville's conclusions about herders in his report on "Grazing in the Forest Reserves," and he added that sheep grazing actually reduced the likelihood of fires by removing flammable vegetation: "In denying the charges of firing the woods, the sheep men correctly point out that the closely fed park lands are less liable to be fired, and that in many cases fires have actually come to an end when reaching closely cropped sheep ranges" (Roth 1902, 347). This idea was common, especially among range livestock producers. Oregon sheep owner John Minto was among the first to publicize it from his position as secretary of the State Board of Horticulture in the 1890s (Rakestraw 1958). A letter to the editor of *American Forestry* by one Aaron W. Frederick (1910) of North Fork, California, expressed it this way: "A well regulated pasturing of the forest land is a necessity. The herbage must be cropped off as a prudential and protective measure; otherwise the grass, drying, makes a great tinder-box to destroy the curtained hills. Horse and cow do a work the ranger cannot. They are among the greatest public benefactors of the national forest."

Prominent foresters were more reluctant to connect the dots between livestock, grass, and fires, however. Henry S. Graves, cofounder of the Yale School of Forestry and Pinchot's successor as chief of the Forest Service, published a bulletin on "Protection of Forests from Fire" in August 1910, shortly after assuming the post and just eight days before the "Big Blow-up" burned more than three million acres in the northern Rockies, throwing the credibility of the young Forest Service into question (Egan 2009). Graves observed that "Nearly all forest fires start as surface fires" and that "The severity of a surface fire depends largely on the quantity of dry material in the forest" (Graves 1910, 8). "In many forests," he went on, "the presence of grass constitutes one of the important problems connected with surface fires. . . . Grass a foot high, if dense, may produce such a hot fire as to start a crown fire" (10). But when he came to the question of "the prevention of fires," Graves made no mention of reducing grasses or other fine surface fuels, whether by livestock grazing or other means. Instead, he inveighed against "the practice in many parts of the South and West, of setting out fires to burn off the litter and brush, usually for the sake of better range" (28).

During the 1910s and 1920s, the possibility that livestock grazing could be a positive tool for preventing fires was actively discussed within the Forest Service, although it only infrequently made it into public reports or publications. In 1912, Assistant District Forester John H. Hatton sent a questionnaire to 160 national forest district supervisors, of whom 120 responded. For reasons that are unclear, the results were not published until 1920, but the conclusions were unequivocal: the average size of fires on ungrazed lands was 273 acres, compared to just 15 acres on "normally stocked" lands and 4 acres on "overgrazed" lands. Eighty-one percent of respondents agreed that grazing reduced the incidence of fires, and more detailed records from seven national forests showed "that grazing causes an average reduction in the number of fires of over 60 per cent" (Hatton 1920, 5). Only "a very small number of stockmen, largely among the more illiterate class of Mexican herders," persisted in the belief that fires were good for forage production, Hatton wrote, and having herders or cowboys out on the range aided in detecting and suppressing fires. "The advantages of having the grazing permittees and their employees on the Forests during the fire season far outweigh the disadvantages" (10).

Hatton's concluding recommendations closely resembled those made by James Jardine in a 1915 lecture at the Yale School of Forestry entitled "Grazing." Jardine noted that grazing reduced fuel loads, and that stock trails and driveways could serve as fire lines and improve firefighters' access to remote areas. Water sources developed for livestock could also aid in fighting fires. Jardine even went so far as to advocate "overgrazing of strategic points where the sacrifice of the small area will aid in the protection of a much larger and more valuable area" (Jardine 1915, 16), and he argued that the benefits of grazing to fire protection justified investments to open up remote forest areas to livestock. "Live stock grazing in the National Forest is an important factor in fire protection. The benefits outweigh the harmful effect and the net gain is an advantage for conservation" (14). Hatton repeated all these recommendations, many of them in identical language, and he also called for research on lightning incidence to help identify the best locations for driveways and "strategic points" "for possible amelioration by grazing." He closed with a recommendation for "the closer correlation of the live-stock industry of the Forests with the fire-protection plans" (Hatton 1920, 11).

As far as I am aware, Hatton's report was the last public admission by the Forest Service that overgrazing might be justified as a means of protecting forests from fire. In his Yale lecture, Jardine had stated: "The ideal use in general is normal grazing. The intensity of grazing that will maintain the pasture at continuous production should be aimed for. Overgrazing is justified along fire lines or driveways where it will result in greater protection of the total forest" (Jardine 1915, 15). But the line between normal grazing and overgrazing was indistinct, and it appears that the latter term was too obviously negative to be openly espoused. Instead, the subject moved out of public view into the internal conferences and memos of Forest Service range officials. At the agency's 1923 Grazing Conference held in Ogden, Utah, the head of the Division of Fire Control, Roy Headley, argued that grazing was the only tool available to contend with "extreme fire danger," such as had occurred in 1910 and 1919: "What can we do about it? The things we can do about it are not very many. We can remove snags, make fire breaks, can adopt measures to keep fires from starting and, perhaps, the most important of all, we can remove inflammable litter or reduce the inflammability of the litter and growth on the ground by grazing. . . . When we have no grazing, as we have unfortunately in some places still, we have in the majurity [*sic*] of instances a pitiful lack of ability to control fire, particularly in these two years" (US Forest Service 1923, 20). Headley called for "full grazing," because "Where there is waste feed there is unnecessary fire danger" (21). When asked by Forest Service Chief W. B. Greeley (who succeeded Graves in 1920), whether he would "deliberately overgraze that Forest in order to get greater safety from fire," Headley replied:

> That is the last thing I would say. The revision of dominating ideas in carrying capacity matters would be the proper thing. I do find that there is 25 to 50 per cent of feed left, when the grazing may [*sic*] would say that 10 per cent is all that needs to be left in the interest of forage production. I'd like to see a recognition of the need of utilization as complete as the grazing man would say is permissible. I'd like to see a strong effort made to get that utilization because of its importance in fire control. Can we afford to go on as I think we are in regarding grazing so lightly in regard to fire; can we afford to think so much of the forage, so much of the fat stock idea, so much of improving the range, so much of improved administration, and so little of benefit of timber protection? (US Forest Service 1923, 22)

That this line of questioning came from the head of fire control, rather than a Forest Service range scientist, is telling. Headley presumed that the scientists knew how much grass could be grazed without harm to the range, but his real concern was the threat of fire to timber, not the threat of overgrazing to forage. For the range scientists, however, whether 10, 25, or 50 percent of the forage in a given year should remain ungrazed was a question to which no one yet had a solid answer—indeed, many already saw it as *the* question they faced, and they would continue to do so for decades to come. Two years earlier, at the Forest Service's District 3 (Arizona and New Mexico) Grazing Studies Conference in Albuquerque, how to estimate carrying capacity had been a dominant topic, whereas fire had not come up at all. The Ogden attendees apparently shared Greeley's and Headley's view that overgrazing should not be permitted, but they avoided specifying what that actually meant. The conference's subcommittee on fire protection, chaired by none other than Aldo Leopold, crafted four resolutions that received unanimous conference approval under the heading "Reducing Fire Hazard by Grazing":

1. There are areas on the National Forests of high fire hazard, where this hazard can be greatly reduced by normal grazing.
2. Distribution of stock to accomplish this should be undertaken in cases where there is now incomplete utilization, making such concessions as may be necessary to stockmen to introduce stock into inaccessible areas.
3. Destructive grazing is not usually required to accomplish the desired purpose; due regard may therefore be given to maintaining the range in productive condition so as to furnish continuous annual range and stable grazing.
4. Where studies or knowledge of local conditions indicate that desired [timber] reproduction would be prevented, forest growth injured, or watershed values impaired by the kind of grazing needed to reduce fire hazard, decision must be reached on the basis of the best available data as to which is most important. (US Forest Service 1923, 70–71)

In the 1920s, terms such as *normal*, *regulated* (Korstian 1921, 279), *careful*, and *moderate* (Clapp 1926, 13–14) grazing were used in public reports when referring to the potentially beneficial effects of livestock for fire prevention. The word *overgrazing* was avoided, and by the 1930s, the idea that grazing was a tool for preventing fires effectively disappeared from public documents altogether, although it still occasionally showed up in internal memos.[1] Why?

The interactions of forage, grazing, and fire defied the bureaucratic separation of range and forestry that was gradually hardening within the Forest Service research enterprise. Apart from Hatton's simple survey, very little research was done on the topic, and it effectively fell through the cracks between the two branches. In his annual report of 1916, Jardine alluded to "minor investigations" under way in California and on the Jornada Range Reserve "to determine the effect of burning the range on the subsequent production of forage," apparently instigated at the behest of stockmen who favored the practice. But the data were not yet in, he wrote—and none was publicly reported thereafter.[2] The following year, he reported that studies using fire to reduce brush in California and Oregon had produced short-term success but were soon followed by the return of brush. From the vantage point of range research, the question was whether fire could be a means of improving forage, not whether grazing was a means of reducing fire.

The foresters, meanwhile, had countless higher priorities for research. Range was only ever a minor part of their agenda, which included silviculture, insects and diseases, forest products and materials, watershed protection, and timber supply economics in addition to fire prevention and suppression. When the Forest Service reorganized its Research Branch in 1915, range was not even included, instead remaining in the Branch of Grazing until 1926 (Godfrey 2013, 75–76). Three major reports suggest that the status of range within Forest Service research grew ever more marginal over time. In a tabular report of the "Forest Service Investigative Program" for 1914, the "Grazing" section took up 9 pages out of 100. Hatton's survey was the only investigation touching on fire, and it was described as for internal administrative purposes only. In 1926, *A National Program of Forest Research* devoted 16 pages out of 232—about 7 percent—to range. By 1933, when the secretary of agriculture transmitted to Congress *A National Plan for American Forestry* range took up just 34 pages out of 1,677—a mere 2 percent.

The *National Program of Forest Research* warrants close attention because it served as the blueprint for the McSweeney-McNary Forest Research Act of 1928, which dramatically increased funding for Forest Service research of all kinds (Godfrey 2013, 108–14). A key provision of the act "designated the Forest Service as the primary agency for federal research on wildland fire. While this move helped secure fire research within the Forest Service, it also discouraged other agencies from pursuing fire studies on their own" (Pyne, Andrews, and Laven 1996, 420), and it did nothing to augment research on rangeland fire. Published by the American Tree Association, the National Program was not an official Forest Service publication, even though its lead author was the head of the Forest Service's Branch of Research, Earle Clapp; rather, it was the report of a special committee of the Washington Section of the Society of American Foresters. The committee's other two members were R. C. Hall, a timber valuation engineer for the Bureau of Internal Revenue, and A. B. Hastings, the assistant state forester of Virginia; their report served both as an argument for augmented federal funding, realized under the McSweeney-McNary Act, and as the basis of the official National Plan seven years later.

As head of the Branch of Research from 1915 to 1935, Clapp "became the principal architect of the Forest Service's nationwide research program, and his views dominated forestry research policy" (Godfrey 2013, 77). In the National Program, he acknowledged that livestock grazing could be compatible with forest management and that the fees paid by livestock producers to graze in the national forests had "so far produced approximately as much income as the sale of timber" (Clapp 1926, 13). Alluding to Hatton's survey, Clapp wrote that "fires spread more slowly and ordinarily do less damage where the area has been grazed. . . . The importance and cost of forest fire control and the obvious relationship of the forage cover to it justify additional and more thorough investigations of the whole problem" (90). Range and forest management were aligned within a common paradigm of maximizing production of the nation's natural resources: "Range perpetuation with maximum grazing parallels forest perpetuation with maximum yields" (89).

Alignment did not mean equality of importance, however; the priority of forests and timber over grasses and forage was unequivocal. "Since the chief purpose in the use of forest land is obviously the growing of timber, it follows that the use of the forage cover must be such that it will, if possible, aid and that in any case it will not interfere with this primary purpose" (Clapp 1926, 87–88). Moreover, this priority was not limited to management but extended to research as well. The "outstanding problems" of range research included "a more exact determination of carrying capacity, including the exact effect of the many factors which influence it, and the perfection of livestock management on the range. *All research, however, whether fundamental or applied, must have constantly in the background the primary requisites of timber growing* and the preservation of the soil conditions which are essential not only to the growing of the maximum timber crops but also to the maximum crops of the forage itself" (Clapp 1926, 92, emphasis added). When the official *National Plan for American Forestry* came out in 1933, it made no mention of grazing in the section on "Protection against Fire." W. R. Chapline, James Jardine's successor, was nearly silent on the topic of fire in the two sections he wrote about rangelands. In the second paragraph under "Forest Ranges," Chapline asserted that grazing "has a direct influence on fire protection" (Secretary of Agriculture 1933, 527), but he said nothing about what that influence was; in a short section entitled "A Forest Range Program," he wrote nothing at all about fire.

Whether one's primary concern was soil, forage, timber, or fire, how to manage grazing appeared, as ever, to hinge on the question of carrying capacity. To return to the words of Leopold's 1923 subcommittee, what was "normal grazing," and what was "destructive"? It was easy to say that grazing should be managed to maintain "the range in productive condition so as to furnish continuous annual range and stable grazing," but how much grazing would actually do that, on the ground, in the various and diverse national forests? How could the effects of livestock grazing on vegetation be understood scientifically, such that predictions could be made and management guidelines formulated? Clapp's National Program did not answer these questions, but it did announce how he expected the Branch of Research to go about finding answers. Under the heading "Ecology," he wrote: "Plant association development and succession of the subordinate forest vegetation [i.e., grasses] has been and is now being studied by many university ecologists in the East, although *most of the field is yet to be covered. . . . Practically the entire question of association development and succession . . . still awaits investigation*. Its importance may be indicated by *a single example*, in which the Forest Service found in central Utah that the climax and highly valuable wheat grass type could be retained only under proper range management. Excessive grazing resulted in lower successional stages of much less value" (Clapp 1926, 84, emphases added).

In an odd rhetorical two-step, Clapp admitted that almost nothing was yet known with any scientific certainty, and then declared that the path to such certainty was clear. He embraced Clementsian ecological theory well before its scientific merits had been firmly established and even while academic ecologists were challenging it. It may have appealed to him and his Forest Service colleagues in part because of its implicit arboreal chauvinism—for Clements, the paradigmatic highest stage of vegetation was a forest—which resonated with the longer history of scientific forestry in Europe (Davis 2016). In any case, the single example on which Clapp based his assertion was the work of Arthur Sampson at the Forest Service's Great Basin Experiment Station in the Wasatch Mountains of Utah. From this slender empirical basis, Clapp would go on to impose successional theory on Forest Service range research, as we will see in the next chapter. Fire was passing into a blind spot for the emerging field of range science, and what occluded it was successional theory and its central concept, climax. To understand how this happened, we must return to the Wallowa Mountains in 1907 and pick up the story of Sampson's career.

Climax

As we learned in chapter 1, Frederick Coville plucked Sampson from the lab of Frederic Clements at the University of Nebraska on his way west to the Wallowa Mountains in the spring of 1907. Sampson was just completing his master's thesis on the "Influence of Physical Factors on Transpiration." Clements was about to leave Nebraska for a position at the University of Minnesota, and presumably he could have recruited Sampson to join him there for doctoral studies. Instead, he recommended him to Coville, and Sampson would not receive his doctorate, from George Washington University, until ten years later.[3] Probably neither the student nor his mentor lamented the parting, however, since they could scarcely have been more different in temperament. No biography of Clements has been written, despite his enormous and somewhat controversial influence on the field of ecology.[4] But it is known that he was "an intensely serious individual. By almost all accounts he was aloof, a bit arrogant, and inflexible in his scientific views. So obsessed was he with his research that as a young scientist he would forget to eat or sleep, a practice that sometimes led to hallucinations. Puritanical in personal habits, he abstained from tobacco and alcohol, and was distressed by colleagues who did not. As Roscoe Pound, the noted jurist and a fellow student at the University of Nebraska, commented wryly, 'Clements has no redeeming vices'" (Hagen 1993, 182). Sampson, by contrast, was famously gregarious both socially and physically. An accomplished distance runner on the Nebraska track and field team, he was also an enthusiastic wrestler and boxer and an avid sports fan. A former student at Berkeley recalled that Sampson "told us he was not going to stand for any 'messing around' in class, but if we so desired he would show us off-campus that he could out-story, out-box, and, if necessary, out-party any of us. A couple of evening tours with Sammy in San Francisco nightspots proved he was not bragging" (Parker et al. 1967, 348).

Getting away from Clements clearly benefited Sampson's career. As Hagen (1993, 186) notes, "None of the scientists who worked with [Clements] found it easy to establish an independent professional or intellectual identity." Sampson suffered no such fate. His results in the Wallowa Mountains were well received, and when the Forest Service founded the Utah Experiment Station in 1912 (renamed the Great Basin Experiment Station in 1918), he was appointed its director, where he stayed until 1922, when he took a faculty position in the Forestry School at the University of California, Berkeley. Widely considered "the father of range science," he wrote the field's first textbook, *Range and Pasture Management* (Sampson 1923), trained countless future government range specialists, and mentored many of the most prominent range scientists of the next generation before retiring in 1951.

Sampson did not part company from Clements intellectually, however. The timing of his training at Nebraska was fortuitous. Clements published

2.1. Frederic Clements (in boots) at the Santa Rita Experimental Range in southern Arizona, 1939. Source: Santa Rita Experimental Range archives.

Research Methods in Ecology in 1905, laying out a suite of techniques and methods developed over the preceding decade while studying the (rapidly disappearing) Nebraska prairies with Pound, Bessey, and others. Sampson got the full benefit of these newly invented methods, which were revolutionizing plant ecology by placing it on a quantitative basis both methodologically and epistemologically (Tobey 1981). Also, an early statement of Clements's theory of plant succession appeared in 1904 as *The Development and Structure of Vegetation*, a volume published by the Botanical Survey of Nebraska, of which Bessey was head. But the full formulation of the theory would not appear until 1916 in Clements's magnum opus, *Plant Succession*, which may have made it easier for Sampson to proceed with his own, parallel investigations over the intervening years. In any event, he never publicly criticized or repudiated his mentor's ideas, and they routinely cited each other's work for decades to come.

Clements's legacy in ecology is, as Hagen (1993, 182) puts it, "ambiguous. Clements's ideas were tremendously controversial even during his lifetime. Although today he is universally recognized as an important intellectual pioneer, many ecologists use him as a scapegoat for supposedly discredited ecological concepts." Foremost among these concepts is the one Clements considered both necessary and fundamental to ecology itself: "the assumption that the unit or climax formation is an organic entity" (Clements 1916, 3). Communities of plants, he believed, were themselves organisms whose stages

2.2. Arthur Sampson circa 1912. Source: USDA Forest Service, Rocky Mountain Research Station.

of development were as fixed as those of an individual plant. Henry Gleason (1926, 1939) forcefully (and repeatedly) challenged this premise, insisting that the community was simply an assemblage of individual plants competing with one another to germinate, grow, and reproduce. Also much criticized, most famously by the early English ecologist Arthur Tansley, was Clements's view that the climax community was determined only by climate interacting with soils, and that the activities of people and animals could only produce "subclimaxes," even if the resulting vegetation patterns were stable for very long periods of time. Tansley (1923, 48) termed such communities "biotic (anthropogenic) climaxes," the first use of the word *anthropogenic* in its current sense.[5] Equally important for present purposes, Clements saw fires in a similar light, even if they were not human-caused. He acknowledged that "Grassland areas are produced the world over as a result of burning and grazing combined, and they persist just as long as burning recurs" (Clements 1916, 108), and he recognized that recurrent fires could keep brush and trees from establishing in some grasslands (including southwestern savannas; Clements 1920, 320). But he referred to these as subclimaxes, climaxes maintained by fire, or fire climaxes rather than true climax communities. In a much more general sense, Clements is associated with the idea that the dynamics of ecological systems are *equilibrial*: feedbacks counteract swings in component parts such that disturbances are naturally absorbed and reversed to maintain a "balance of nature." Stability, rather than change, is therefore the natural state of things.

Although Clements was hardly alone in embracing the balance of nature (Wu and Loucks 1995), and he does not really deserve credit or blame in the more recent emergence of non- or disequilibrium ecological theory (chapter 7), both Gleason and Tansley were right, at least in principle and especially in retrospect. Few scientists today would dispute the strong influence of humans on vegetation everywhere, whether mediated by fire, livestock, or anthropogenic climate change, and there is growing evidence of this influence deep into the past (Mann 2006; Ruddiman 2005). Carl Sauer (1950) anticipated these ideas in (of all places) the *Journal of Range Management*, where he challenged the climax concept by arguing that grasslands worldwide had coevolved with humans' mastery of fire for hundreds of thousands of years. But the idea and ideal of pristine, original nature remains tenacious in many people's minds, especially among US environmentalists, and the belief that "nature" will restore itself if "protected" from human activities is widespread. Similarly, although Gleason was ridiculed at the time, most ecologists since the 1950s have shared his "individualistic" conception of ecological succession. But the matter is not as simple as being right or wrong. As we will see, something like Clements's ideas of climax, equilibrium, and organismic communities was indeed necessary to make ecology into an applied science, at least on US rangelands.

In his first report on his research in the Wallowa Mountains, Sampson wrote unequivocally: "The methods of study followed were those developed by Dr. F. E. Clements" (Sampson 1908, 13). He divided the Wallowas into four zones, based on the climate at different elevations and the dominant plant species found in apparently undisturbed sites. He then demarcated "five typical stations representing large tracts of overgrazed range land" in three of the zones and had the stations fenced to keep livestock out (Sampson 1908, 12). Within each station, Sampson installed numerous one-meter square quadrats—the core of the techniques that Clements had pioneered in the Great Plains. He permanently marked the quadrats with stakes at each corner, photographed them, and mapped the location, size, and identity of every plant. He then denuded some quadrats of all plants and removed the top three or four inches of soil to monitor the invasion or recolonization of plants following disturbance—what Clements termed *secondary succession*. Over the course of the summer, Sampson observed the timing of germination and flowering of every plant in the quadrats, and he took measurements of soil moisture and air humidity; air, soil, and surface temperatures; and light intensity. He also collected seeds and planted them under controlled conditions to determine rates of viability and germination, and he watched sheep grazing on adjacent areas to ascertain which plants they preferred to eat and when. He repeated all these procedures during a second season in the summer of 1908 (Sampson 1909).

As Tobey (1981, 68) notes, "The invention of a quantitative method for ecology was more than the clever application of statistics. The invention of the quadrat, or meter-plot, embodied a profound epistemological shift, in which the scientists ceased to believe in the reality of one phenomenon and began to believe in the reality of another phenomenon." Clements and his colleagues in Nebraska showed that general or "impressionistic" observations of vegetation could be misleading; accurate knowledge required enumerating each and every plant in quadrats and analyzing the data statistically. The epistemological authority of Sampson's training and methods was a major reason Coville had hired him, and the Wallowa quadrats were meticulous and rigorous in this fashion. His assigned task, in Coville's words, "was to ascertain, by a detailed study of range vegetation, whether by closing an area for a portion of the season the best grazing plants could not be made to reseed and sufficient time be left after the seeding was accomplished to permit the sheep to enter the area and graze off the season's growth of vegetation" (Sampson 1908, 6). Many sheep owners in the area were proposing the complete closure of overgrazed rangelands for a period of years to allow recovery of the forage, but Coville sought a way to avoid this if at all possible for the sake of the local grazing economy.

What emerged from Sampson's observations soon came to be known as "deferred rotation grazing," a strategy that still remains a staple of Forest

Service range management recommendations. Sampson found that almost all the highest value forage plants were perennial species that only rarely reproduced from seed even under controlled conditions, instead propagating vegetatively from established individuals. The annual plants were much less preferred by sheep, and they reliably produced abundant seed even in the presence of grazing. Most importantly, the high-value forage plants set seed by September 1, roughly four-fifths of the way through the growing season, meaning that they could then be grazed without harm to either the existing plants or the prospects for future establishment. This would be "equally effective" as complete closure in securing range improvement, and perhaps even better, "since the stock would do the work of a harrow by trampling most of the seeds beneath the surface of the soil, and thereby insure a higher percentage of germination" (Sampson 1908, 20).

In his early reports, Sampson suggested that only overgrazed areas be targeted for deferred grazing, and he cautioned that it might take several years for significant improvement to occur. But after two more summers of research in the Wallowas, and a much more sophisticated analysis of the data from his quadrats and other experiments, he extrapolated his findings into a strategy for general application: by *deferring* grazing of a portion of a range—for example, one-fifth—and *rotating* the deferred portion over a period of years—for example, five—the whole range would be assured of a chance to establish new high-value perennial forage plants. He hadn't studied fire directly, but Sampson nonetheless wrote that one of the three advantages of deferred grazing over both yearlong grazing and complete protection was "the removal of the vegetation itself, thus minimizing the fire danger from an accumulation of flammable material" (Sampson 1914, 143).

From a more theoretical perspective, the key contribution of Sampson's Wallowa research was to shift the interpretation of grazing impacts from the physical destruction of present biomass to the ecological alteration of future species composition. For Jardine and the Coyote-Proof Pasture Experiment, the question was how to optimize the use of the current forage crop by reducing the amount lost to trampling, thereby increasing the yield of wool and mutton per animal and per unit of land. For Sampson, by contrast, the question was how grazing altered the dynamics of succession, changing the kinds of plants that would grow in the future. "Deferred grazing is based upon the requirements of the vegetation through practically a double life cycle" (Sampson 1914, 125)—from the germination of a plant through the germination and establishment of its offspring. From this perspective, trampling could be a means of helping one year's seeds become the next year's new plants. Neither Jardine nor Sampson directly addressed these contrasts in their reports, but the potential to understand, predict, and control future range conditions nevertheless made Sampson's work valuable for the Forest Service and important to the future of range science.

Grazing was not the primary reason for the creation of the Utah Experiment Station in 1912. Located on the western slope of the Wasatch Plateau in the Manti National Forest south of Provo, Utah, it was established to study watershed protection following severe flooding that threatened downstream communities, dams, and irrigation works. About 4,600 acres in size, it ranged from 6,800 to 10,400 feet above sea level; average annual precipitation graded with elevation from 12 and 30 inches. Thus, it was much smaller, higher, and wetter than the Santa Rita and Jornada Experimental Ranges (chapter 3).

When Sampson was appointed the first director of the Utah station, he was given a variety of scientific tasks in addition to overseeing a small staff and the construction of facilities. His third annual report, from December 1914, described forty-one silvicultural projects, as compared to just eleven under "Grazing Studies"—and one of the eleven was simply "Climatic Observations." To study the watershed, Sampson oversaw installation of a sediment basin, measured the surrounding vegetation, and demonstrated that grass recovery dramatically reduced flooding and sedimentation—a finding that was cited frequently in subsequent Forest Service reports and publications (Sampson and Weyl 1918). He also devoted considerable time to the study of artificial reseeding of the range using both native and imported plants. But Sampson's most lasting and influential research concerned "natural reseeding" and how it was affected by livestock grazing. As in the Wallowa Mountains before, he pursued this topic through the lens of successional theory, informed by data from scores of quadrats. These studies culminated in the publication of USDA Bulletin number 791, *Plant Succession in Relation to Range Management*, in August 1919.

Bulletin 791 was the most influential article in range science for the next thirty years, the period in which the field coalesced into a discipline; range scientists did not significantly challenge its core ideas for more than half a century (chapter 7). "The carrying capacity of a large portion of the millions of acres of western range has been materially decreased by too early grazing, overstocking, and other faulty management," Sampson began. "Stockmen generally recognize this fact and are doing what they can to overcome these faults in management and to increase the productivity of the range. Where grazing has been subject to regulation for some years and the stock has been handled according to the most approved methods the productivity of the range has been appreciably increased." He then explained the key problem that the bulletin would solve:

> One of the most serious drawbacks in the past has been the lack of a means of recognizing overgrazing in its early stages. In deciding upon the lands especially in need of improvement, the stockmen and those regulating grazing have essentially relied upon general observations of the abundance and luxuriance of the forage supply and upon the condition of the stock grazed. The depletion of the lands is seldom recognized by these general observations until their carrying capacity has been materially reduced, or until the animals grazed are in poor condition of flesh. So long as the cover is more or less intact, there is little indication that the range is being slowly but certainly depleted; the depletion is not recognized until the more palatable and important forage species are in low vigor, and their growth and reproduction seriously impaired, or perhaps not until a large proportion of the plants actually have been killed. Until there is insufficient feed to support the animals, they will retain their condition of flesh fairly well; but long before there is insufficient feed to satisfy their appetites a large portion of the vegetation is killed. (Sampson 1919, 1–2)

Succession provided a means of defining, detecting, and avoiding overgrazing by looking not at the density of forage or the condition of the livestock but at the composition of the plant community. In a diagram, Sampson arranged five "states of development" of vegetation and soils from algae and lichens on bare rock upward to deeply rooted perennial grasses on rich, moist soil (figure 2.3). These states reflected "the laws underlying the occupation of lands by vegetation" (Sampson 1919, 2), and succession would impel any site toward the "final or climax stage" in the absence of disturbance. This was a straightforward adaptation of Clementsian theory to rangelands. Sampson's key addition was to conceive of grazing as pushing directly back against succession in a linear fashion:

> In the utilization of lands as grazing areas, the invasion by the higher type of vegetation is often prevented, especially where the species high in the development are grazed with greater relish than those lower in the succession. Thus the plants well up in the development of the type may disappear gradually or suddenly, according to the degree of disturbance caused by the adverse factor, until the plant stages lower in the development predominate. If the factor adverse to the progressive development of the vegetation continues to have its play for an indefinite period the vegetation will continue to revert until

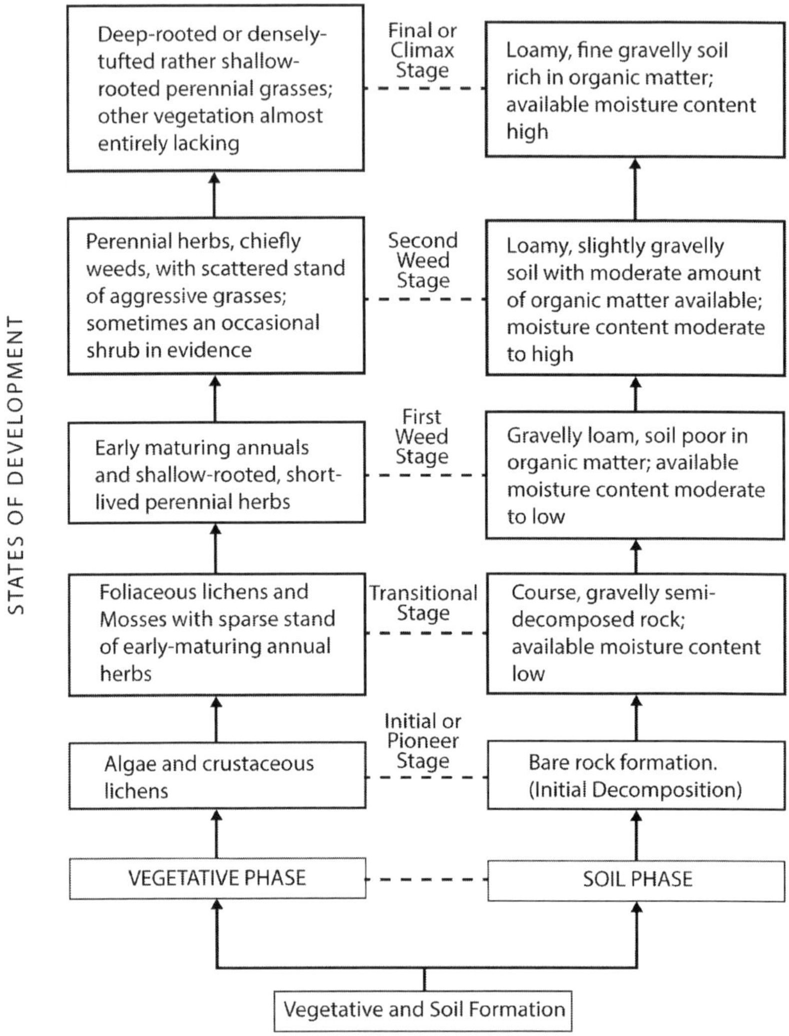

STATES OF DEVELOPMENT

Deep-rooted or densely-tufted rather shallow-rooted perennial grasses; other vegetation almost entirely lacking
Final or Climax Stage
Loamy, fine gravelly soil rich in organic matter; available moisture content high

Perennial herbs, chiefly weeds, with scattered stand of aggressive grasses; sometimes an occasional shrub in evidence
Second Weed Stage
Loamy, slightly gravelly soil with moderate amount of organic matter available; moisture content moderate to high

Early maturing annuals and shallow-rooted, short-lived perennial herbs
First Weed Stage
Gravelly loam, soil poor in organic matter; available moisture content moderate to low

Foliaceous lichens and Mosses with sparse stand of early-maturing annual herbs
Transitional Stage
Course, gravelly semi-decomposed rock; available moisture content low

Algae and crustaceous lichens
Initial or Pioneer Stage
Bare rock formation. (Initial Decomposition)

VEGETATIVE PHASE
SOIL PHASE

Vegetative and Soil Formation

2.3. Arthur Sampson's conceptual diagram of rangeland plant succession. Redrawn from Sampson (1919, 4) by Darin Jensen.

the first-weed stage reappears, or, indeed, until practically all the soil is carried away and the pioneer stage returns. Such a succession of the plant cover down the scale from the more complex to the primitive type will be referred to in this bulletin as retrogression, retrogressive succession, or degeneration. (Sampson 1919, 5)

Livestock grazing could now be understood as *the* independent variable determining rangeland vegetation. Other factors, such as roads, trails, or prairie dogs, could cause retrogression in discrete sites, Sampson admitted, but "overgrazing or other faulty management is usually accountable for the retrogression in the vegetation on range lands as a whole" (Sampson 1919, 6). Sampson acknowledged in a footnote that the concept of retrogressive succession differed from Clements's theory, which reserved the term *succession* for the upward or forward development of plant communities. He further allowed that retrogression might not occur "in the same specific descending series as it has been recorded to occur in the ascending development toward the climax"—which may explain why the arrows in his diagram only pointed up. As if to placate his mentor, Sampson insisted that "'retrogression' or 'retrogressive succession' is a convenient and self-explanatory term, and its use in no way involves a fundamental principle" (Sampson 1919, 5). Indeed, the coherence of retrogression required the coherence of succession as its opposite.

Livestock did not necessarily cause retrogression, however. Echoing his findings from the Wallowas, Sampson stressed that well-managed grazing did not inhibit progressive succession, and might even promote it by aiding germination (Sampson 1919, 54). Most plants could withstand early-season grazing once every three or four years, so deferred rotation could ensure that "the vegetation will retain its vitality almost as well as when not utilized at all, provided, of course, that the number of stock carried is correctly estimated" (Sampson 1919, 64). What Sampson offered was a way to distinguish overgrazing from normal grazing: if retrogression of the plant community was occurring, then grazing was excessive and/or poorly managed, even if the overall amount of vegetation had not declined. Species from lower successional stages were "plant indicators," to use the name of Clements's 1920 companion volume to *Plant Succession*. By measuring the plant composition, and comparing it to the successional sequence and climax, in theory one could detect adverse change in its early stages, while there was still time to do something about it.

In theory. Sampson did not report actual stocking rates in his Utah experiments, and he gathered examples of retrogression primarily by observing the vegetation in heavily grazed areas outside the experiment station or previously disturbed sites such as stock trails, bedding grounds, or around watering points. What his famous bulletin did not do was specify the actual amount of grazing—whether as a percentage of the forage crop that was consumed (subsequently termed *utilization*), the number of livestock per unit area, or the length of time they were there—that would trigger retrogression in a given plant community. Obviously this would depend on a site's vegetation, which Clementsian theory linked to two relatively stable factors: climate and soil. But it would also depend on the weather, and especially on rainfall, which could vary widely from year to year. Sampson clearly recognized this: he documented divergent monthly rainfall totals from year to year, and he carefully measured the influence of soil moisture on forage growth. But he did not venture to explain how variability in the weather should affect livestock management.

Clements, too, was acutely aware of variability, and he had more to say about it than Sampson. *Plant Indicators,* which was the applied companion volume to the more theoretical *Plant Succession,* both echoed and cited Sampson's work. "The difference in the total yield of the same range in two successive years of dissimilar rainfall may be greater than 100 per cent," Clements noted, and "grass types show a carrying capacity cycle of excess and deficit, which must be taken into account if alternate lack of utilization and overgrazing are to be avoided" (Clements 1920, 292–93). He employed and discussed the concept of carrying capacity, but Clements explicitly rejected averages as guidelines for management. "It is evident that the maximum production [of forage] can not have a fixed or average value. . . . A degree of grazing which would be disastrous in a drought period would fall far short of adequate utilization during a wet one" (Clements 1920, 296). He believed that this variability was itself natural, and he expressed confidence that sunspot cycles—a topic of widespread scientific enthusiasm at the time—were the underlying cause. Thus, even dramatic swings in forage production were ultimately evidence of equilibrium.

Indeed, temporal variability was central to Clements's recommendations for the management of livestock and his conviction that damaged rangelands could always be recovered. He reconciled contradictory early accounts of prairie vegetation by arguing that no snapshot description could be taken as conclusive. If some observers saw short grasses and others tall ones at the same location, it was due to the transient effects of both weather and bison. Clements reasoned that drought periods had inevitably resulted in overgrazing in the Great Plains, prompting the bison herds to depart for greener pastures elsewhere. Thus: "An enforced period of rest ensues, during which successional processes bring about the restoration of the original grass cover, unless again disturbed by overgrazing. It is this cycle which, in its beginning, has been especially disastrous to the grazing industry of the West, just as the subsequent and inevitable regeneration through succession offers the solution of all overgrazing problems. . . . This is not an excuse for overgrazing, . . . but it does make clear that *all overgrazed ranges can be certainly and greatly improved by proper rest or rotation. This is the basis of all range improvement*" (Clements 1920, 310, emphasis added).

For both Sampson and Clements, deferral and rotation were based on the insight that "rest"—the removal of grazing—for key periods of plant growth would maintain and improve forage by encouraging recovery and succession. Clements (1920, 296) asserted that following livestock removal, "the successional process begins and soon terminates in the original climax"; that rest would cause restoration was "universal and inevitable in all climaxes." But the question of carrying capacities and proper stocking rates remained unanswered for practical purposes; all that could be said with confidence was that both were variable over time.

James Jardine, as head of grazing research for the Forest Service, immediately recognized the value of Sampson's research for avoiding degradation and enabling recovery of already degraded rangelands. In his 1919 annual report he wrote: "Investigations at the Great Basin Experiment Station have developed facts which indicate the possibility of determining minor changes in the improvement or deterioration of ranges through comparatively slight changes in the plant composition. If this means of detecting changes in range condition proves reliable and practicable upon wider study it will be an important advancement. . . . If the new principles and methods work out in practice it should be possible to tell whether a range is improving or deteriorating before any marked damage takes place and consequently the necessary measures to effect improvement will influence grazing capacity but slightly" (Jardine 1919, 5–6). From Jardine's perspective, the further value of *Plant Succession in Relation to Range Management* was precisely the opposite of what Clements counseled in *Plant Indicators*. That is, it would enable livestock owners and the Forest Service to *avoid* disruptive adjustments to stocking rates: "prompt detection of deterioration will be imperative to the maintenance of the ranges *without recurring changes in the numbers of stock*" (Jardine 1919, 6, emphasis added). Chapline reiterated this view two years later at a Forest Service conference on grazing in the Southwest: "Investigations have developed facts which show that minor changes in the improvement or deterioration of ranges can be determined through comparatively slight changes in the plant composition. The immense importance to grazing administration of this investigation is the fact that range deterioration can be checked in its early stages by slight adjustments in grazing management. This will prevent overstocking with resultant cuts in number of stock grazed, which is always a difficult matter to handle where grazing conditions are as intensive as they now are on most National Forests" (US Forest Service 1922, 131). This interpretation—or misinterpretation—must be counted as one of the most important moments in the history of range science. Jardine knew very well that forage production varied widely, especially in the Southwest: only five years earlier, he had recommended that stocking rates on the Jornada Range Reserve be adjustable by at least 20 percent, as we will see in the next chapter. But when scientific knowledge conflicted with the interests of livestock owners and agency administrators, the latter took priority.

If Clementsian *theory* became foundational to range science via Sampson, then, Clements's *practical* insights faded into obscurity via Jardine and Chapline. Unlike bison on the Great Plains of yore, sheep and cattle in fenced allotments could not simply migrate away from overgrazed or drought-stricken ranges and allow rest and succession to restore them. Deferred rotation could mimic migration, but it would prove difficult to implement without interior fences—and even the perimeter fences of most allotments had yet to be built. Finally, in place of varying stocking rates to match dynamic carrying capacities, range management and administration on the national forests would soon write stable stocking rates (called "preferences") into the terms of each allotment's lease. Jardine's curious logic—that if retrogressive succession could be detected early, no significant stocking adjustments would be necessary—conflated what Sampson's work over the preceding decade had distinguished to such powerful effect: the production of biomass from the composition of vegetation. Two temporal scales were interlinked: production varied in the short term, with and because of variations in rainfall, and this determined the likelihood that a given amount of grazing would trigger long-term retrogression in species composition. But stocking adjustments were difficult to make quickly, for ranchers and the Forest Service alike. The greater the variability of rainfall in a given place, the greater the problem this would pose to range scientists.

The prevailing ideas and structures established for rangeland administration and management by the early 1930s were persistent. Fire became the bureaucratic monopoly of the Forest Service's Division of Fire Control, separated from range research and effectively taboo for range scientists within the agency for the next half century. Clementsian climax theory became the seldom-questioned basis for understanding and studying rangelands throughout the West. Carrying capacities and stocking rates—how many cattle or sheep grazed how much land for how many months—became the central issue for scientists and administrators alike, who shared an abiding faith that rangelands would restore themselves "naturally" by succession if livestock were reduced, rotated, or removed. The legendary battles between agencies and ranchers, and more recently between ranchers and environmentalists, all hinged on these flawed ideas about how rangelands work.

The problems that lurked in the blind spots of the emerging field of range science were not invisible to everyone, however. As we will see in the next chapter, scientists in drier regions, particularly the Southwest, glimpsed them with increasing frequency, and some tried to bring them to the attention of the Forest Service in internal reports. Aldo Leopold (1924, 6) even published these words in the *Journal of Forestry*: "Until very recently we have administered the southern Arizona Forests on the assumption that while overgrazing was bad for erosion, fire was worse, and that therefore we must keep the brush hazard grazed down to the extent necessary to prevent serious fires. In making this assumption we have accepted the traditional theory as to the place of fire and forests in erosion, and rejected the plain story written on the face of Nature." Why were these concerns not heard, and why did they have so little effect on rangeland science and administration more generally?

The answer lies in the larger political and economic context surrounding the use of the national forests and other public rangelands. Static carrying capacities and stable stocking rates, buttressed by Clementsian theory, served both bureaucratic and financial interests by focusing on a controllable variable that could be stipulated in leases and enforced on the ground (albeit with some difficulty). Even more importantly, leases with fixed stocking rates—*X* number of sheep or cattle for *Y* number of months each year, presumptively renewable in ten-year increments—could be collateralized by ranchers and bankers relatively easily. As Laurence Stoddart (1935, 531), one of the foremost midcentury range scientists, observed: "The commercial value of a range is primarily determined by the number of livestock it is capable of supporting. A method of measuring this capacity is, therefore, highly desirable as a key to value." Fixed carrying capacities meant that publicly owned rangelands could be privately capitalized on the basis of the market value of the livestock that the leased allotments promised (at least on paper) to produce, facilitating both credit and the market exchange of the ranches that held leases. Scientists and administrators alike recognized this advantage from the outset and expressed it in both publications and internal reports.[6]

Although the evidence is only circumstantial, what went unspoken may have been equally decisive in winning the assent of the Forest Service as a whole: static stocking rates in highly variable and drought-prone regions dovetailed perfectly with the overriding goal of fire suppression. Grazing the same number of animals in both wet and dry years ensured heavy grazing of fine fuels at precisely the times when severe fires were most likely—that is, during droughts. "Well before systematic fire control, cattle and sheep cropped fire from the land, and they did so with a thoroughness that engine companies, smokejumpers, and helitack crews have never equaled."[7] Scientists and agency officials recognized this internally until 1920, but they never acknowledged it publicly, apparently to avoid the appearance of enshrining overgrazing in policy.

Instead, fire virtually disappeared from the Forest Service's range research agenda, just as grazing remained invisible to forestry research, even as livestock did the work of helping to prevent fires throughout the West for most of the twentieth century. Leopold transferred from the Southwest to Wisconsin in 1924, and he left the Forest Service altogether in 1933. When Sampson addressed the topic of fire at the Inter-American Conference on the Conservation of Renewable Natural Resources (1948, 551), he allowed for the possibility that "*controlled* range burning may be economically sound under some conditions" but characterized research on the subject as "limited," and most of the studies he cited came from state experiment stations and universities rather than the Forest Service. In 1955, when the Society of American Foresters published a review of *Forestry and Related Research in North America*, its discussion of fire prevention said nothing about grazing, and its discussion of range research said nothing about fire (Kaufert and Cummings 1955). Until about 1980, setting fires for experimental purposes was nearly impossible within the agency, even in places with little or no commercial timber and even when fire appeared economically advantageous.

Two apparent exceptions merely serve to prove the rule. A pair of Forest Service scientists burned 6,000 acres of dense sagebrush rangelands in southeastern Idaho in the 1930s and concluded that "planned burning" could be a means of reducing sagebrush and increasing forage production. They went to great lengths to contrast planned burning with "accidental or haphazard burning," however, which "nearly always produces damage or loss, often of disastrous proportions," and they advised that other, mechanical methods "should be used in preference to burning" if possible (Pechanec and Stewart 1944, 9, 3). By the 1950s, mechanical and chemical methods would displace fire as a sagebrush management tool altogether.[8]

Robert Humphrey worked for the Forest Service and the Soil Conservation Service from 1933 to 1948.[9] Something of an iconoclast, he was the only range scientist, according to West (2003, 504), who "formally disputed" the Clementsian idea of climax in the 1940s, and he pursued fire throughout his career despite significant resistance. Humphrey conducted one small experimental fire in 1933, although he had to do it outside the Forest Service's

Santa Rita Experimental Range on a private ranch nearby, and he did not publish the results until after he left federal employment for a position at the University of Arizona (Humphrey 1949).[10] By the late 1950s, Humphrey had data showing that fire could reduce shrubs and cacti on southwestern rangelands; "the combined evidence appears conclusive," he wrote, "that grassland fires in the desert grassland, as perhaps in grassland areas the world over, have been instrumental in preventing the establishment of woody species" (Humphrey 1958, 242). He was ahead of his time, however, and his research program struggled for funding after an experimental fire on the Prescott National Forest escaped control in 1955 (Carle 2002, 64). (Other Santa Rita scientists conducted a handful of experiments in the 1950s to test fire effects on shrubs and grasses, but the results were deemed unsatisfactory, as we will see in chapter 5.)

Of the few other studies that managed to reach publication between 1930 and 1970, several resulted from accidental or wild fires that happened to burn in sites where data had been collected for other purposes (e.g., Cable 1965; Dwyer and Pieper 1967). Even in such sites, fire could disappear into the blind spot created with Earle Clapp's premature judgments of 1926. In an article published in *Ecology* in 1945, for example, Walter Cottam and Frederick Evans (1945) of the University of Utah compared two canyons in the Wasatch Mountains above Salt Lake City (north of the Great Basin Experiment Station) from 1935 to 1944. They attributed the differences they observed—higher perennial grass cover and lower shrub cover in the ungrazed canyon—entirely to livestock grazing, even though the ungrazed canyon had been regularly burned by the military "to clear the land for target ranges." Without addressing any role that fire might have played, they inferred that grassland was the "pristine" condition of Great Basin foothills "at the time of the first white settlement," and that livestock alone were responsible for subsequent shrub encroachment. That the article passed scholarly peer review suggests that the fire-climax blind spot extended even beyond the Forest Service.

Today, scientific studies, reinforced by increasingly frequent catastrophic forest fires throughout the West, have thoroughly discredited the fire suppression policies that shaped twentieth-century federal land management. The ecological changes that have resulted from a century of fire suppression, however, coupled with the persistent bureaucratic grip of past policies and procedures, have made reintroducing fire on any large scale virtually impossible. In many higher-elevation forests there is so much accumulated fuel that fires seem unsafe even under carefully prescribed conditions. Southwestern rangelands have the opposite problem: shrubs have replaced the grasses that once provided fine fuels for fire spread, and where such fuels do exist they are often invasive nonnative grasses that thrive with burning. Meanwhile, declining forage production squeezes the economic viability of ranching, making alternative land uses (e.g., subdivision and residential development) more likely, which in turn makes fire still more difficult to restore.

How to evaluate the effects of grazing and fire on western rangelands is a confounding question to this day, since one cannot easily control them separately: wherever there is grass, grazing suppresses fires, while not grazing makes fire almost inevitable. The more scientists study fire, the more complex it appears to be: in the Southwest, for example, burning in winter has different results than burning in spring or summer, and fire effects may depend less on the characteristics of the fire itself than on subsequent rainfall or drought. Nonetheless, it is clear that in many parts of the West fire is necessary for the maintenance of desirable conditions for people, wildlife, and watersheds alike. Yet fire restoration has occurred on only a few, discrete locations (Sayre 2005b). The best conditions for burning are, generally speaking, those that match the evolutionary norm—when drought follows a period of heavier rains—but that is precisely when the risks are highest and resources (equipment, firefighters, etc.) are in greatest demand from wildfires. Whether the steady decline in livestock grazing on public lands—which has occurred throughout the West but especially in the vicinity of growing suburban areas—has contributed to increasing fire size and frequency, is a question that has not yet been studied. Meanwhile, even though fighting wildfires now consumes more than half of the Forest Service's total budget, prescribed fire is still considered too dangerous and unpredictable, it seems, to be countenanced as a management tool for widespread use. The blind spot has been exposed, but most people still choose to avert their eyes.

THREE

Squinting at Blind Spots

Southwestern Rangelands and the Consolidation of Successional Theory

The mismatch between successional theory and US rangelands was inversely related to precipitation. Although all the lands in question were dry relative to the eastern United States, there was considerable variation among rangelands in different parts of the West in terms of both average annual precipitation and interannual variability. Relatively moist conditions prevailed in the prairies of Nebraska, where Clements developed his theory, and the same was true at the higher elevations, where Sampson did his most seminal research and where most national forests were located. Interannual variability was correspondingly lower, generally speaking. Leaving aside the question of fire, succession captured the dynamics of vegetation in relation to range livestock in such places reasonably well. But in lower and drier rangelands, successional theory and the concept of carrying capacity foundered. Evidence of this was detected from the earliest days of government range research, but it would take the better part of a century to be openly acknowledged, and still longer for its full implications to be appreciated (chapter 7).

It was in the desert grasslands of the Southwest that the blind spots of range science created the most conspicuous problems, in part because that is where so much of the earliest research was conducted. Forest Service histories of range science (e.g., Shapiro 2014) often begin with Sampson's work at the Great Basin Experiment Station, which was the first of its kind to be established by the Forest Service. But the Bureau of Plant Industry (BPI) had already created two others: the Santa Rita Range Reserve near Tucson, Arizona, in 1903, and the Jornada Range Reserve outside Las Cruces, New Mexico, in 1912 (map 3.1). Both were transferred to the Forest Service in 1915 and renamed experimental ranges in the 1920s; until the 1930s, the Great Basin, Santa Rita, and Jornada were the primary (and sometimes the only) sites of federal range research. Thanks to Sampson, the Great Basin Station is generally

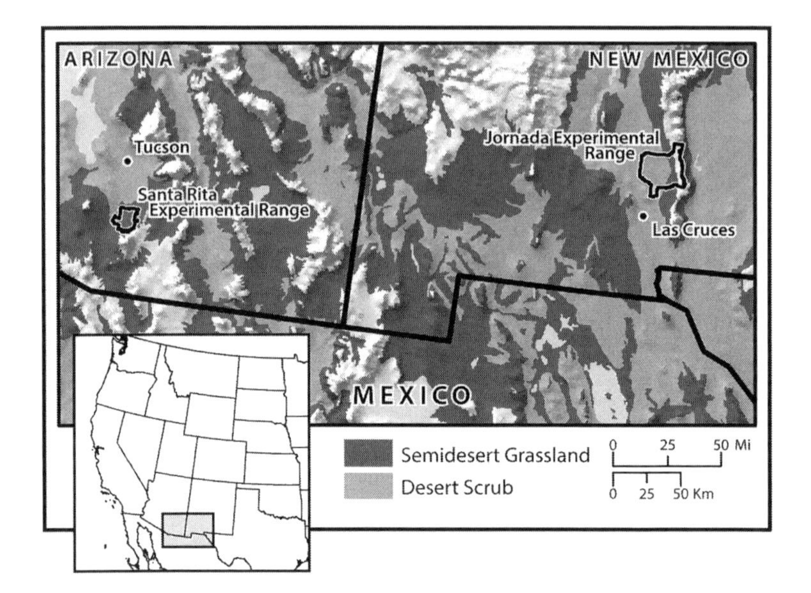

Map 3.1. Major biotic communities of the US–Mexico borderlands, showing the locations of the Santa Rita and Jornada Experimental Ranges. The excluded (white) areas include both high-elevation mountains and forests and low-elevation deserts. Significant portions of Sonoran and Chihuahuan Desert scrub area, including much of the western two-thirds of the Jornada, were previously semidesert grassland. Source: adapted from Brown and Lowe (1980) by Darin Jensen and Alicia Cowart.

credited with having had a greater formative influence on US range science, but in terms of the overall history of the field, the Santa Rita and Jornada ranges would ultimately be of greater significance; their relative obscurity may be due to the fact that research at both places so often disappointed officials in the Forest Service. The reason for their disappointment, however, was that southwestern rangelands did not conform to the scientific, bureaucratic, and economic expectations that consolidated around Clementsian theory. The questions the scientists were given could not be successfully answered on the terms and time lines imposed on them. The value of the research they did would not become apparent until decades later, in part because a long temporal scale was needed to reveal the dynamics of more arid rangelands.

The Santa Rita and Jornada Range Reserves

Both the Santa Rita and the Jornada range reserves were founded expressly to conduct research at scales large enough to be relevant to actual ranches in the Southwest. In 1900, the BPI had arranged for 336 acres just outside of Tucson to be reserved from homestead claims by the secretary of agriculture and fenced to keep livestock out. David Griffiths, employed first by the Division of Agrostology and then by the BPI to study Arizona rangelands, detected little sign of recovery over the following three years (Griffiths 1904). Drought conditions from 1898 to 1904 contributed to this finding, but the tract was also deemed simply too small. "The production of forage is so small here, at best, that one is obliged to measure his pasture by square miles rather than by acres, and the operations in range improvement must be on a correspondingly large scale" (Griffiths 1904, 29). In 1903, a much larger area—31,488 acres, or almost 50 square miles—of the recently created Santa Rita Forest Reserve, about 30 miles south of Tucson, was fenced by permission of the Department of Interior (which still administered the Forest Reserves) and set aside for range investigations (figure 3.1); an executive order in 1910 expanded the area to 65 square miles. Somewhat higher than the earlier tract, the new reserve spanned an elevation gradient from 2,900 to 4,500 feet above sea level, and average annual rainfall graded with elevation from about 10 inches to nearly 20. As such, it contained a more representative sample of Arizona's semidesert grasslands, from lower sites with mostly annual grasses and forbs to higher ones dominated by perennial grasses. Shrubs, especially mesquite, were scattered across the reserve and thicker along drainages, and, like the rest of the region, the entire area had been severely overgrazed during periodic droughts since the 1880s.

From the very beginning, the focus of research was squarely on the question of carrying capacity. Griffiths set out 28 plots spanning the new reserve, each three-by-seven feet, and measured the growth of all herbaceous vegetation in both the winter and summer growing seasons. Based on these numbers, he calculated that the range produced an average of 350 pounds per acre of usable livestock forage per year. Assuming that an adult cow needed 18 pounds of feed per day, he concluded that 37 acres were needed to support one animal for a year. At the time, this was probably the most sophisticated carrying capacity measurement yet made on US rangelands. It appeared to answer the question that Clements, Sampson, and the Forest Service range experts would avoid when they convened for conferences in the 1920s. But Griffiths himself was cautious, if not altogether skeptical, about his estimates. The new reserve was not being grazed at this time, and therefore might not accurately represent the carrying capacity of adjacent ranges. The bigger problem, however, was variability. Griffiths observed "a wide variation in the quantity of vegetation which is produced even in areas situated near each other" (1904, 30), and he was acutely aware of the complex relationship between rainfall and plant growth. Precipitation varied widely from year to year and

3.1. The fenced boundary of the Santa Rita Range Reserve in 1905. The photo was taken by David Griffiths to illustrate the contrast between grazed lands, to the left of the fence, and lands protected from grazing for two years. Source: National Archives, Forest Service photograph 402893.

place to place, not only in quantity but also in its timing relative to the growing seasons of different plants (60). "There is without doubt no part of the country where the character of the native feed is so variable as it is in the Southwest" (46).

Unlike so many other influential early range scientists, Griffiths had not studied at the University of Nebraska (he earned his PhD from Columbia in 1900), and he made no use of succession or climax in his writings. He was clearly aware of Clementsian ideas, however, and he used the Great Plains as a kind of baseline to evaluate what he saw in the Southwest, drawing strong contrasts both in the quantity of forage produced and in underlying vegetation dynamics. "Before any rational adjustment for the proper control of public grazing lands" could be made, he wrote, "much should be definitely known regarding the amount of stock that these lands will carry profitably year after year. This must form the basis of all equitable allotments. To secure such information is a most difficult task in a region where the seasons, the altitude, the slope, and the rainfall are so variable. It can be determined very easily in the Great Plains region, where conditions are uniform and reasonably constant, and indeed it is very definitely known there; but here the case is very different" (Griffiths 1904, 32). After six more years of observations and experiments, Griffiths concluded that "the natural restocking of the perennial range by new plants takes place at irregular intervals," when good summer rains happened to occur in two consecutive years: one to produce seed, the second for germination (Griffiths 1910, 12). The weather, rather than livestock, appeared to be the main driver of rangeland dynamics. It was exceedingly hard to discern any pattern in those dynamics, moreover, especially among the annual plants. Even in the absence of grazing, "differences in vegetation, comparing one year with another, are very striking. . . . In the large field, even with similar rainfall, there occurs an ascendency of one plant one year and another plant another year. . . . So far as known, no one has ever offered an explanation for these yearly variations of annual vegetation" (Griffiths 1910, 15). Interestingly, Clements (1916, 1920) would later cite Griffiths's work, but Sampson made no mention of it in his influential 1919 bulletin that adapted plant succession to range management.

Experiences at the Santa Rita led directly to the founding of the Jornada Range Reserve in south-central New Mexico as an even larger site for rangeland research. Elmer O. Wooton worked for Griffiths as an assistant beginning in 1911, and one of his tasks was to write up the latest results of the research at the Santa Rita. According to a letter he wrote in 1936, Wooton remarked to Griffiths "that the best data in the records were the figures showing the 'animal-days' feed furnished to cooperators' stock by known areas of land" (a notably finer resolution than the animal unit months that would later become the standard measure of forage on public rangelands). The cooperators were livestock owners who had agreed to participate in research on three fenced pastures adjacent to the most productive, upper edges of the Santa Rita. The goal was to observe the effects of controlled grazing, and significant improvement had occurred since the pastures had been fenced in 1908 (Griffiths 1910, 10). Griffiths agreed with Wooton, but he worried that the data were from areas that were unrepresentative and too small (less than 600 acres) to be applied more broadly; he believed that larger-scale studies were needed. But the BPI had no money to build sufficient fencing. (The Santa Rita cooperators had paid for their pasture fences themselves.) Wooton had worked previously in southern New Mexico, and he knew of a rancher there named Charles Travis Turney, who had built up a ranch north of Las Cruces by purchasing scattered 40-acre homesteads and grazing some 5,000 cattle on the public domain surrounding them. Wooton convened a meeting with Turney and the director of the New Mexico Agricultural Experiment Station, Luther Foster, where Turney agreed to pay to fence the land where his animals grazed and to provide the scientists data about his management, if they could get him legal permission to do so—fencing of the public domain still being illegal at the time. Wooton recalled:

> Our principal desire was to obtain accurate data on grazing capacity on a large area. We expected to let Mr. Turney run the ranch as he would do if it were his own land, and compare the results with those on ranches on the open range from which we could get data. Mr. Turney knew more in five minutes about managing a cattle ranch than I would ever know. We all thought we knew that proper management of stock and care of the feed would result in improvements of many kinds, and we wanted to prove the correctness of our faith by seeing it occur and keeping as accurate records as possible of the progress and rate of improvement.[1]

On May 3, 1912, President Taft signed an executive order setting aside 192,434 acres as the Jornada Range Reserve.[2] Wooton served as the first director, and Turney continued to run livestock there in cooperation with the researchers until 1925.

Both the Santa Rita and the Jornada remained in the BPI at this time, but it wasn't long before the apparent redundancy of two agencies studying the same topic came to the attention of USDA officials in Washington. In June 1914, B. T. Galloway, acting assistant secretary and former chief of the BPI, sent a memo to the current BPI head, William Taylor, and copied it to Alfred Potter, chief of grazing for the Forest Service. "The work that Doctor Griffiths and Mr. Wooton are doing in connection with the study of public range would appear to be rather anomalous. . . . The Forest Service now has a well-organized branch for the study of range conditions, and some excellent work is being done by this branch of the service. It is a question whether it would not be advantageous for all concerned if all this range work could be handled by the Forest Service."[3] Galloway closed by informing Taylor that a special committee, consisting of Taylor, Potter, and Frederick Coville, would be tasked with considering "all questions relating to the range work now being conducted by the Department to the end of uniformity of action and the securing of results which will tend to aid in the development of a general policy for the administration of public lands."

Five months later, James Jardine conducted a nine-day examination of the Jornada, apparently at the behest of the special committee. His report concluded that the reserve should be retained by the USDA precisely for the study of rangelands *outside* the national forests, lands which the Forest Service did not manage or administer but which Congress had nonetheless instructed the agency to include in its range research efforts. Jardine saw the Jornada's location, size, and aridity as advantageous in this connection: "The Jornada Reserve is representative of a large body of range lands in Arizona, New Mexico, and Texas, which will perhaps remain purely range land longer than any other large body of what are at present range lands, due to general lack of moisture for cultivated crops and occasional severe droughts." From a practical standpoint, the extreme marginality of the land on the Jornada meant that long-term, large-scale experiments could be undertaken with little risk of "conflicting with settlement," and working with a single cooperator, such as Turney, would be easier than working with several. Interestingly, though, Jardine also offered a more scientific rationale: "The main principle worked out on the Jornada Reserve, for the building up, maintenance, and management of range will be applicable to a greater degree under more favorable conditions, than results obtained under more favorable conditions would be to the large body of desert range of which the Jornada Reserve is representative." This may have been little more than a hunch on Jardine's part, but it is worth emphasizing in retrospect: It was wiser to extrapolate findings from more marginal conditions to less marginal ones than the other way around—precisely the opposite of what would later happen with Sampson's results from the Great Basin Experiment Station. In 1915, the USDA transferred responsibility for all range research, including management of the Santa Rita and Jornada, from the BPI to the Forest Service.

In his report, Jardine repeatedly emphasized the importance of research to determine the carrying capacity of southwestern rangelands. Efforts on the Jornada needed to be much more aggressive, he believed, including comprehensive use of quadrats to measure vegetation in carefully delineated "range types" throughout the reserve. The actual current carrying capacity, compared to the potential capacity under improved management, provided the organizing conceptual framework for research. But the overarching rationale for the reserve turned on fencing. The surrounding public domain was unfenced, open to any and all livestock owners who chose to graze their animals there, and the resulting lack of control was universally regarded within the USDA as the obvious cause of western rangeland destruction everywhere outside of the national forests. Jardine recommended modifying the contract with Turney to emphasize the contrast between "controlled" and "uncontrolled" grazing in relation to livestock mortality (especially during drought), utilization, calf crops, carrying capacity, and operating costs, all "with a view of benefiting stockmen."

At both the Jornada and the Santa Rita, research was expected not only to produce scientific knowledge about how rangelands worked but to demonstrate the practical value of scientifically informed management practices. Fencing was seen as a prerequisite to both: control of animals was necessary for scientists to draw robust conclusions about grazing as much as it was for ranchers to improve their lands and their economic returns. And the best way to convince ranchers of the wisdom of fencing was to show them by example, not simply with data from quadrats but at the scale of actual ranches. The scientists were effectively running two large demonstration ranches. From the 1910s into the 1930s, the annual reports from both range reserves shared a common refrain: controlled, scientific range management produced more grass, more calves, larger animals, and larger profits than uncontrolled grazing on the open range. As in the Wallowa Mountains before, however, such comparisons necessarily relied on fragmentary and potentially incomparable data from surrounding, unfenced rangelands.

The extreme variability that had puzzled Griffiths now produced a tension between "fundamental" scientific inquiry and practical demonstration. To avoid large livestock losses during droughts, Jardine's 1914 report urged that roughly half the forage on the Jornada be reserved for use during the driest months and that deferred grazing be used in areas of perennial grasses that would retain their nutrients when "cured on the stalk" during the annual spring dry season. Calves were the primary product of southwestern ranches, but Jardine recommended that about 20 percent of the Jornada herd consist of "steers [castrated males] for ready sale or shipment" rather than breeding cows. Assistant Forester Will Barnes went further in a memo the following year: about one-third of the herd should be steers, to enable rapid sale "to relieve the range in case of prolonged drought without interfering with the breeding herd." In his 1917 annual report, Jardine called for research into "what reduction in grazing capacity and changes in grazing management are necessary during drought and how rapidly the stockmen can figure on building up their herds after a drought is broken. These investigations are fundamental in livestock production on the ranges of the Southwest."

Frequent adjustment of stocking rates was not only difficult for ranchers, however; it also posed challenges for scientific research. Jardine acknowledged this in his internal reports: "Perhaps no phase of grazing work is more important than the matter of grazing capacity. This subject, however, is difficult to attack by controlled experiments." How could scientists test the effects of various stocking rates if those rates were constantly changing? Something other than animals and area—the numerator and denominator of carrying capacity—would be needed. Measuring utilization—the percentage of a given year's forage crop that was grazed—emerged as one possible solution, but this was also difficult. Chapline's 1921 annual report remarked: "One of our biggest problems is to work out just what is proper utilization and satisfactory methods of determining this for technical range inspection. This line of investigations seems the most promising of satisfactory results." His optimism was premature: utilization is intuitively coherent and can be qualitatively useful, but *quantifying* it would confound agencies and scientists for decades to come (Smith et al. 2005), in part because "grazing itself may induce additional new growth of herbage" (Scarnecchia 1999, 158).

The Santa Rita typically received more precipitation than the Jornada, but researchers there reached similar conclusions. A 1926 internal report noted that "the forage crop in a good year may be as much as five times that of a poor one, and the good years number about one in five, the balance ranging from poor to fair. Full appreciation of this fact is absolutely essential to the proper estimation of carrying capacity" (Culley and Hill 1926, 73). Beginning in 1915, scientists at the Santa Rita had cooperated with ranchers to graze cattle on the reserve, and researchers used rainfall data, actual stocking rates, and measurements of the density and productivity of major forage species to calculate carrying capacities for three range types in terms of acres per animal. They found wide variation from year to year: as much as 62 percent on the "foothill" type and nearly 100 percent on the "mesa" type. Multiyear averages could be computed, of course; they "cannot, however, be expected to meet the conditions brought about by the periodic severe drouths that occur in the southwestern range country" (Culley and Hill 1926, 95).

Observations such as these did not go unnoticed in Washington. *Range Management on the National Forests* for example, noted that forage production in a given place could vary by an order of magnitude from one year to another (Jardine and Anderson 1919, 24). Chapline observed that in the Southwest, "rainfall in individual years may fall to one-third of average and the stand of forage, even without grazing, may drop 80% or more from its maximum. . . . Starvation losses of from 30 to 50% of the livestock may occur if management is not properly adjusted" (Chapline 1933, 5). Variability was not only temporal but also spatial: rain gauges on the Jornada showed 25 to 50 percent variation in the same year between locations only six to twelve miles apart (US Forest Service 1922, 92). Awareness of such variability may help explain why Jardine and Chapline were so excited by the apparent promise of Sampson's work to enable stable stocking rates.

Until the late 1920s, the Santa Rita and Jornada reserves were considered largely successful as demonstrations of improved management. Jardine reported in 1917 that three years of "light grazing" during the summer growing season on the Jornada had improved 47,000 acres of grama grass range by at least 50 percent over adjoining outside range (cf. Jardine and Hurtt 1917, 6). His annual reports repeatedly cited higher calf crops on the reserves, compared to those of surrounding ranchers' herds, as evidence of the superiority of controlled grazing strategies, in particular deferred rotation. Chapline echoed in his 1921 annual report: "The losses from all causes on the [Santa Rita and Jornada] Reserves have been only about one-third to one-fifth of the average losses on outside ranges. Calf crops have been fifty per cent greater, and in some cases have been double what have been secured on the outside ranges" (Chapline 1921, 10–11). He boasted that "the investigations at these two Reserves conclusively show that livestock production on the uncertain, semi-desert southwestern ranges can be stabilized" (Chapline 1921, 5).

The findings were even more impressive because they occurred in the face of periodic dry spells, while the open range in the surrounding region continued to deteriorate. The Jornada experienced poor summer rains from 1916 to 1918, drastically reducing perennial grass density, but the range recovered with good rains in 1919 and 1920 (US Forest Service 1922, 93–94). This was followed by another drought from 1921 to 1925, which cut forage production by half; two years of better summer rains restored those losses on the reserve but not on the adjacent open range (Chapline 1928, 6–7). In a 1933 address, Chapline (1933, 14) praised Jornada grazing management for securing "as good range conditions for 16 years as has total protection from grazing." On the Santa Rita, likewise, he cited experiments showing that "a semi-desert grama range can be maintained in good condition as well, or better, with proper grazing than it can with total protection from grazing" (Chapline 1928, 7). Grazing, he explained, "tends to keep the clumps of grasses smaller, does not allow such an accumulation of dead dry grass, and prevents over-development so that when a dry year comes the vegetation on the ground appears to be able to withstand the smaller amount of soil moisture available" (Chapline 1928, 3).

The reports, publications, and memos generated by scientists working on the Santa Rita and Jornada Experimental Ranges in the first three decades of the twentieth century do not fit the conventional understanding of the rise of range science. Griffiths, Wooton, and their successors observed extreme variability in rainfall—spatially and temporally, at both seasonal and interannual scales—and they noticed the variety of impacts that livestock grazing could have in interaction with that variability. In some cases, those impacts were both severe and long lasting, as had been abundantly demonstrated during the cattle boom of the late nineteenth century. But under controlled conditions, the effects of grazing were inconsistent, and in some ways negligible or even beneficial compared to complete livestock exclusion (Culley and Hill 1926). Even as they worked diligently to measure forage and calculate carrying capacities, they counseled against static stocking rates, reasoning that the variability in rainfall—and therefore forage—required a corresponding flexibility in the number of livestock grazed on a given piece of rangeland. Moreover, these scientists did not study and think about southwestern rangelands in terms of succession and climax, even if they used tools (such as quadrats) and concepts (such as range types and carrying capacity) that Clements had also used. Griffiths, for example, recognized that the removal of fire for an extended period—itself a result of the cattle boom—could alter vegetation dynamics in ways that were self-reinforcing and therefore potentially permanent (chapter 5).

Problems Revealed and Occluded

The McSweeney-McNary Forest Research Act of 1928, modeled on Earle Clapp's 1926 *National Program of Forest Research*, finally answered the long-repeated pleas of Jardine and Chapline for augmented range research funding. The budget for the Jornada grew from $45,000 in 1928 to $123,000 in 1933, and funding for range research overall increased nearly eightfold, from $36,000 in 1925 to $275,000 in 1938 (Ares 1974, 30). The number of full-time range researchers grew from fewer than 40 nationwide in 1927 to roughly 200 twenty years later (Campbell 1948, 6). But the overall effect was to diminish the prominence of the Santa Rita and the Jornada, both because new experimental ranges were founded elsewhere—in Utah (1933), California (1934), Colorado (1937), and Oregon (1940) (Chapline et al. 1944)—and because under the statute range research became less autonomous from Forest Service research as a whole. The Office of Grazing Studies had been transferred from the Branch of Grazing to the Branch of Research, under Clapp's direction, in 1926, and in 1930, Clapp reorganized Forest Service research into ten regional offices, with the five western regions each combining range and forestry. The new arrangement was supposed to enable greater integration of forestry, range, economics, and watershed research. But rather than being two out of three research sites reporting to a Washington office specifically concerned with grazing questions, the Santa Rita and Jornada now became two of five units within the new Southwestern Forest and Range Experiment Station, headquartered in Tucson. The effects of these changes were evident by 1939, when a two-week Range Research Seminar was convened at the Great Basin Station. Only twenty-six of the fifty-five seminar participants held range-related positions, and six of those twenty-six had combined titles (e.g., "wildlife and range management" or "forest and range influences"). No one from the Jornada and Santa Rita Experimental Ranges was in attendance, and southwestern range research findings no longer dominated the agenda or discussions, as they had in 1921. Range research was the focus of the seminar, but the larger context was now predominantly forestry. (I will examine the 1939 seminar in more detail in chapter 4.)

During this same period, moreover, the successful demonstration of scientifically informed management at both reserves began to falter, especially at the Jornada. Despite flexible stocking rates, reserve pastures, and deferred grazing, the 1932 Jornada Management Plan reported that the forage base of the reserve had declined nearly 72 percent between 1917 and 1930. Most of the decline could be found in data from 1923 as well, but the recovery of grasses during the wetter years of 1919–1920 and 1925–1926 had not recurred, and losses once considered transient now appeared to be permanent. Whereas Turney had once grazed upward of 5,000 cattle on the same range, scientists now estimated the "sustainable" capacity of the Jornada at just over 1,300 head. Black grama grass (*Bouteloua eriopoda*), the most important forage species in the region, was giving way to shrubs such as tarbush (*Flourensia cernua*), creosote (*Larrea tridentata*), and especially mesquite (*Prosopis glandulosa*). Quadrat measurements of vegetation in areas protected from grazing and grazed at various intensities showed that precipitation—especially during the twelve months preceding a given summer growing season—dramatically affected black grama growth and density, and that heavy grazing could eliminate black grama entirely, especially during a dry summer. More conservative grazing, however, was a minor factor compared to rainfall (Nelson 1934). Subsequent analysis, published in 1965, would show that the transition had begun much earlier: between 1858 and 1915, open grassland had declined from 58 to 25 percent of the reserve. The trend would also continue: by 1963, open grassland was completely gone (Buffington and Herbel 1965, 151).

Rather than demonstrating success, the Jornada now appeared to demonstrate failure, and Chapline found himself forced to defend the reserve's continued place in Forest Service research. In July 1935, he wrote a letter labeled "Confidential" to Arthur Upson, director of the Southwestern Forest and Range Experiment Station, discussing "considerable criticism of the Jornada and of other features of the southwestern range research both inside the Station and outside." He paraphrased the views of David Shoemaker, the head of range management in Region 3, who had complained that, "We are falling down on management. . . . We are not getting at carrying capacity as fully as we should, particularly because of the fluctuations of numbers of livestock. . . . The size of the outfit is probably larger than necessary." Shoemaker also felt "that many of the features which we claim to have developed, stockmen themselves have developed, and we have simply studied and proved the effectiveness of their efforts." (Ironically, this is exactly what Wooton had envisioned for the Jornada at its creation.) "Nearly everyone who comes to the Jornada from mountain range lands or other areas of good production," Chapline complained, "is very much shocked with the scant growth normally found. In a drouth period conditions, of course, appear even worse." But Chapline was skeptical of many of the criticisms, attributing them to "inadequate appreciation of Jornada conditions and a failure to see or understand essential experimental results," and he concluded by informing Upson that funding for the Santa Rita and Jornada would be increased in the 1937 budget.

The problems at the Santa Rita were less abrupt and conspicuous, but mesquite and other shrubs were invading and displacing perennial grasslands there as well. It was a regional phenomenon. In 1936–1937, the Southwestern Forest and Range Experiment Station, the Arizona Agricultural Experiment Station, the Agricultural Adjustment Administration, and the Bureau of Agricultural Economics collaborated to assess "The Occurrence of Shrubs on Range Areas in Southeastern Arizona" (Upson, Cribbs, and Stanley 1937). Researchers made vegetation measurements at 450 sites and conducted visual surveys of some 12 million acres. Grasses dominated nearly one-third of the area, and creosote dominated one-quarter; cactus, burroweed (*Isocoma tenuisecta*), and mesquite (*Prosopis velutina*, a different subspecies than on the Jornada) dominated 7 to 10 percent each; wolfberry (*Lycium pallidum*), saltbush (*Atriplex canescens*), and snakeweed (*Gutierrezia sarothrae*) dominated the remaining 10 percent. Creosote was considered to reflect soil conditions and was relatively stable. But mesquite, snakeweed, and burroweed were singled out as having expanded in recent memory, usually at the expense of grasses; the report attributed this to grazing, without mentioning fire. Up to this point, burroweed had received more attention from Santa Rita researchers than the other two species, but the survey found mesquite to be the most widespread of the three, present on more than 9 million acres or three-quarters of the region. Understanding, explaining, and remedying this shift would be the dominant research priority of Santa Rita range scientists for the next three decades (Sayre 2003).

At both the Jornada and the Santa Rita, Forest Service officials viewed the shift from grass to shrub dominance as a threat not only to regional forage and livestock production but to the credibility of scientific range management as a whole. They did not see it as a challenge to successional theory and the idea of climax plant communities, however, apparently because they were not even thinking in those terms. For all the attention paid to carrying capacities and the need to restore depleted rangelands in the Southwest, the archives of the Santa Rita, the Jornada, and the Office of Grazing Studies through the 1920s are silent on the topic of succession. The perennial grasslands known to have prevailed prior to heavy livestock grazing served as the benchmark and goal for research at both reserves, and scientists noted the tendency for annual grasses and shrubs to colonize sites where perennials had been eradicated—a phenomenon they could have described as retrogressive succession. They also observed that reducing or excluding grazing could help perennial grasses recover, at least when rainfall was sufficient—elsewhere considered a classic instance of secondary succession. But they did not describe these changes in Clementsian terms, nor did they presume that complete protection from grazing would necessarily reverse the damage of overgrazing. The minutes of the District 3 Grazing Studies Conference in 1921, for example, contain no references to succession or climax, and just one passing use of the term *ecological association*. In summarizing the findings from the Santa Rita at the conference, Grazing Examiner R. R. Hill stated:

> (1) Principal grasses of southern Arizona do not require total protection in order to produce maximum amount of forage; and in fact a certain amount of grazing increases the density of stand and the volume of growth. . . .
>
> (3) Heavy grazing, so called, for the year as a whole, tends to maintain the density and even to increase the volume of forage produced by several of the important grasses of southern Arizona. . . .
>
> (4) The greatest production of forage in the case of most of our grasses occurs when the plants are grazed moderately at intervals during the growing season, rather than when the plants are ungrazed until after maturity. This may be called selective grazing as compared with deferred grazing as ordinarily understood.
>
> (5) There is a natural shifting in density and composition of the principal grass species in southern Arizona that is taking place more or less independently of the grazing factor. (US Forest Service 1922, 83)

No one at the conference pointed out that these conclusions ran counter to the successional framework of Sampson's famous bulletin. As late as 1928, Chapline saw no need in his annual report to reconcile successional theory with his observation that controlled grazing was better than complete protection for grassland persistence in the Southwest.

Only in the 1930s did Clementsian terms and concepts begin to appear in the reports and publications of scientists at the Jornada and the Santa Rita (e.g., Campbell 1931; Nelson 1934; Upson, Cribbs, and Stanley 1937). This timing—after the reorganization of Forest Service research in 1930—suggests that successional theory was brought to bear on southwestern range research not at the behest of Jardine, Chapline, or the scientists themselves, but rather under the influence of Earle Clapp, who had endorsed successional theory in his 1926 *National Program of Forest Research,* and who four years later had secured both the authority and the resources to implement that plan. The earliest studies of succession in relation to range conditions for several other parts of the West also appeared in the 1930s (National Research Council 1994, 60), as Clapp opened new facilities as part of his expanding research program. It was the height of the period that Tobey (1981, 117, 121) identifies—from 1917 to 1941—when Clementsianism held unrivaled sway over academic grassland ecology in the United States, with nearly half of all doctorates nationwide awarded at the University of Nebraska alone.

The response of southwestern range scientists was quixotic. Superficially, they began to employ Clementsian theory to organize and interpret their research, albeit with some circumspection. Campbell (1931, 1049) presented data from the Jornada in successional terms, although he noted that in depleted clay soils "succession may require many years or even decades to complete." Nelson (1934, 7) described black grama as "the dominant plant of the climax plant formations" in large parts of New Mexico, but his paper mostly demonstrated the overriding importance of variable rainfall. In the regional assessment of shrub occurrence, Upson, Cribbs, and Stanley (1937, 10) wrote that, "Creosotebush and blackbush [*sic*] are probably climax types," since they had apparently not changed since settlement and grazing. This meant that there was no reason to hope that these areas could be made to grow grass. Areas infested with burroweed, snakeweed, and mesquite, on the other hand, had previously supported grasses, or at least more than at present. "If this be true, then these three types may also be considered, ecologically, a climax grassland type. This, in turn, means that the problem of restoring these three types again to grassland productiveness should offer considerable prospect of success" (Upson, Cribbs, and Stanley 1937, 12). These were the areas that would be targeted for artificial reseeding using machinery and chemicals after World War II, because secondary succession still had not occurred (chapter 5).

On a deeper level, however, reconciling successional theory with observations of southwestern rangelands remained difficult at best. In a mimeographed Forest Service manual entitled *Southwestern Range Ecology,* issued in 1941, three range scientists remarked: "In a region of wide variation in climate, particularly precipitation, it is difficult to determine what is climax vegetation and what is not" (McGinnies, Parker, and Glendening 1941, 112). Plant communities could be identified, but arranging them in a linear succession was sometimes impossible. "Two general types of successional changes are usually recognized, namely: (1) Toward the climax, (2) away from the climax. It is also possible that successional changes may take place which are horizontal or neutral; that is, they are neither toward nor away from the climax" (114). For example, "in Arizona between elevations of 4,000 and 7,000 feet there is a tendency for the development of several types of vegetation; namely, grassland, chaparral, oak woodland, and pinon-juniper woodland" (113). Observations such as these prompted the authors to reject Clements's organic ontology, pointing out that "in a region of diverse topographic and climatic conditions considerable warping and twisting is required to make it fit local conditions" (107). "It should be kept in mind that when we consider plant succession on the basis of plant communities and we recognize certain stages in their development the names applied to these stages are a matter of convenience. Actually succession is a continuous process, and it does not take place on a community basis. The invasion and establishment of the various species are carried on more or less independently, although certain mass effects of vegetation should be recognized" (121).

But even if successional theory was wrong, the scientists admitted that something like it was prerequisite to any range science: "as recognition of changes in vegetation is of primary importance in range management, we cannot escape the implications of plant succession" (McGinnies, Parker, and Glendening 1941, 109). Gleason's individualistic alternative, "if carried too far results in a complete unsystemization as far as plant communities are concerned, and it does not give sufficient weight to the influence of biotic factors" (107). Unable to resolve the problem, the authors of *Southwestern Range Ecology* retreated to the forage-based (and therefore livestock-based) metric of their predecessors, concluding that "progressive succession" occurs when forage productivity increases, whereas "retrogressive succession" is reduced productivity "because of loss of cover, soil depletion, increased run-off or some other reason." Also like their predecessors, they attributed vegetation change to livestock: "Retrogressive succession commonly occurs on overgrazed ranges" (114). Eight pages later, as though recognizing the possible flaws of this argument, they remarked in passing that "there is some evidence to indicate that slow hidden changes take place, so that even where a range appears to be maintained in good condition it is apt to break rapidly during droughts" (122).

Southwestern range scientists found themselves in a contradictory situation. Successional theory was endorsed by their superiors and reasonably well suited to (some) rangelands elsewhere in the country, including many of the national forests, which were the primary concern of the agency for which they worked. The theory had multiple advantages for the Forest Service in relation to administrative efficiency, politically and economically important constituencies, and the overriding objectives of fire suppression and timber production (chapter 2). It also provided a framework that appeared necessary to attain for range research the status of a systematic and predictive science, as opposed to a collection of discrete observations and loose rules of thumb. And the confidence voiced by Clements that range improvement was always possible, thanks to the universal force of succession, provided grounds for optimism that range problems could, indeed, be solved. On the other hand, the scientists had a growing set of empirical observations that appeared to violate the theory, including those already mentioned and more that emerged soon thereafter: evidence from plots on both the Santa Rita and the Jornada that shrub invasion proceeded with or without cattle, sometimes faster in ungrazed than grazed plots (Nelson 1934; Glendening 1952), and from other southwestern sites such as the Desert Lab near Tucson, where fifty years of livestock exclusion did not produce differences in vegetation composition compared to adjacent, grazed areas (Blydenstein et al. 1957).

The Santa Rita and Jornada Experimental Ranges were the oldest range research sites in the world, with more long-term, large-scale data than anywhere else. The scientists there studied and measured the soils and plants and livestock, pored over their transect data, and examined the available records of rainfall and temperature. They described what they saw and tried to make sense of it in as rigorous and coherent a fashion as they could. But when they tried to see southwestern rangelands through the lens of successional theory, they had to squint, and they could not see through the blind spots inherited from other sites and earlier research. If they had had greater autonomy either from forestry and the Forest Service or, after 1930, from the enlarged research apparatus that sprawled across the western United States, perhaps they could have made their voices heard more effectively or demanded more time to develop a theory that matched what they saw. Instead, they reverted to estimating carrying capacities by averaging forage production over time and deducting a portion that should remain ungrazed for the good of the range (figure 3.2). Setting stocking rates in this way ensured both that the range would be severely grazed in dry years and that forage would exceed the needs of livestock in wet years; moreover, *thinking* in these terms reinforced the idea that livestock numbers, rather than rainfall, was the causal factor behind these dynamics.

In summary, scientists at the Santa Rita and Jornada Range Reserves observed from the beginning that extreme variability in precipitation made the idea of stable carrying capacities chimerical. Through the 1920s, they instead

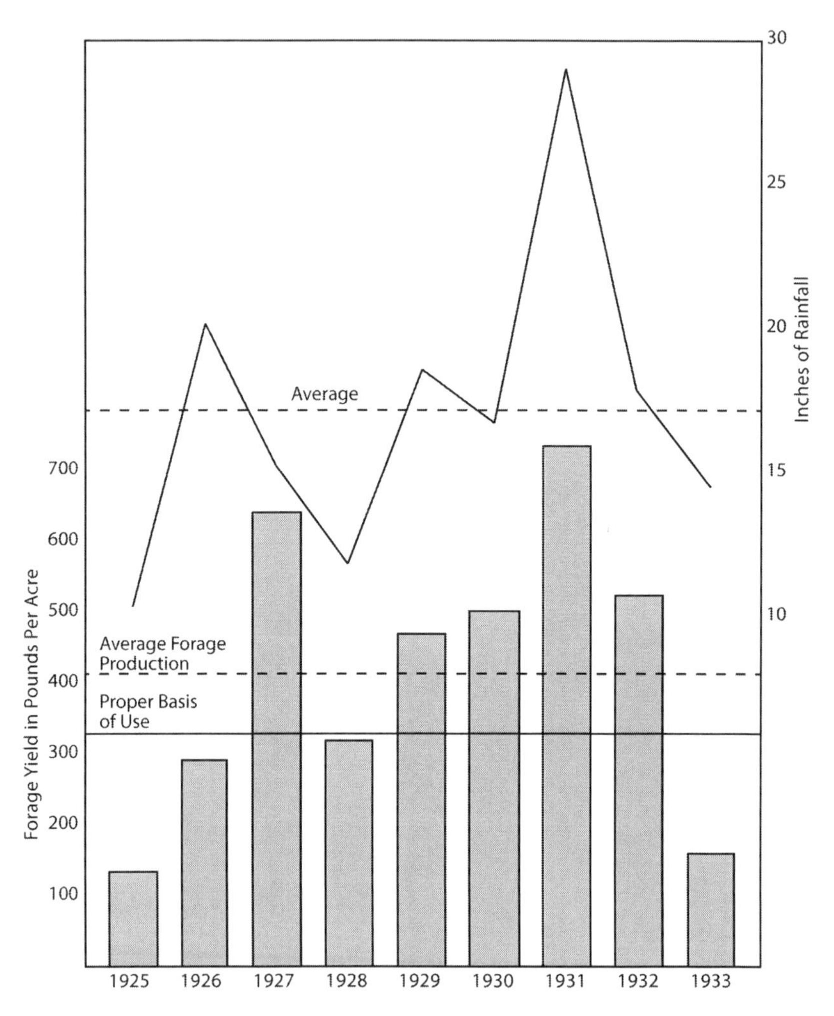

3.2. Rainfall, forage production, and recommended "basis of use"—that is, the amount that should be used to calculate carrying capacity—for semidesert grassland range, based on data from the Santa Rita Experimental Range. Note that stocking on this basis would result in severe overgrazing in two years out of nine (1925 and 1933) and significant excess forage in five years (especially 1927 and 1931). Redrawn from US Forest Service (1952) by Alicia Cowart.

managed to sustain grasslands by varying stocking rates by 30 percent or more every year, even though this ran counter to prevailing ideas in the Forest Service and also made it more difficult to conduct controlled experiments and interpret the resulting data. Above all, they approached the challenge of range science and management without relying on Clementsian theory as the foundation for their ideas. In the 1930s, while Earle Clapp's reorganization of Forest Service research was imposing successional theory as the basis of all range research, threshold changes that defied successional explanations became manifest in the Southwest. Controlling livestock numbers was supposed to have allowed succession to restore depleted grasslands, but vast areas had instead shifted to shrub dominance, some even in the complete absence of livestock. Rather than a single climax community, these rangelands displayed multiple possible stable states, and grazing appeared to be less important in driving change than successional theory predicted. Unable to propose a coherent alternative framework, researchers would instead launch decades of studies designed to kill shrubs and reestablish grasses through other means. Those efforts largely failed as well, as we will see in chapter 5. But eventually, the data the scientists collected at the Santa Rita and Jornada Experimental Ranges would contribute to the development of an alternative model of range science (chapter 7).

The absence of fire is now considered the underlying factor that made shrub encroachment possible (and perhaps inevitable) in the Southwest, but that possibility was scarcely voiced until the 1950s, and scientists at the Santa Rita and the Jornada instead focused their attention on carrying capacities and stocking rates. These were the problems that Jardine and Chapline considered the highest priorities for both research and administration of national forest rangelands. They were problems of monitoring and measurement—simple in principle but remarkably difficult in practice, as we will see in chapter 4. Here the key point is that Clementsian ecology, with its combination of quadrats to measure vegetation and an elaborate classification system of range types, appeared to offer a simple, linear solution to this problem: Measurements made at small scales (quadrats, plots, or transects) could be multiplied by the area of a pasture or range type, and short-term data (collected in one or a series of years) could be interpreted in successional terms and extrapolated forward in time. A relatively wet year, in principle, would counteract the effects of a drier one, just as grazing fewer animals would repair the damage of grazing too many. History did not matter, because climax communities and the successional forces that shaped them were universal attributes of any given site.

From 1915 on, the fact that the Santa Rita and Jornada Experimental Ranges contained no marketable timber had raised recurrent questions about their relevance to the Forest Service, and the agency eventually gave them both up. The Jornada was transferred to the USDA's Agricultural Research Service when that agency was formed in 1953. In its new bureaucratic home, the Jornada continued to concentrate on range science until the 1970s and 1980s, when its scope began to broaden into ecological research more generally as part of the International Biological Program (chapter 7) and the National Science Foundation's Long-Term Ecological Research program. Today, with a staff of some seventy scientists and a new building (named after Elmer Wooton) on the campus of New Mexico State University in Las Cruces, the Jornada is at the forefront of research into the post-Clementsian, nonequilibrium model of rangeland ecology. The Santa Rita remained part of the Forest Service's research enterprise until 1988, when it was transferred to the Arizona State Land Department as part of a large land swap (Sayre 2002); since then it has been leased and managed by the University of Arizona. Research output peaked in the 1970s, with scientists studying range reseeding and grazing management in addition to brush control and, to a limited extent, fire (Martin 1966, 1973, 1978, 1983; Martin and Reynolds 1973; Martin and Cable 1973; McClaran, Ffolliott, and Edminster 2003). As at the Jornada, research at the Santa Rita broadened into ecology more generally, albeit with more limited resources. It continues to be grazed under a cooperative agreement with a local ranching family, and it remains a site for long-term data collection and scientific experiments.

In 1943, Laurence Stoddart and Arthur Smith published *Range Management,* the first range science textbook since Sampson's *Range and Pasture Management* in 1923. For the next forty years (through subsequent editions in 1955 and 1975), "Stoddart and Smith" (as it came to be known) was the standard text for academic instruction in the field, even more so than Sampson's *Range Management: Principles and Practices,* which appeared in 1952 (and was not updated until 1989, by other authors). The appearance of both books at this time reflected the growing number of range management programs at land-grant universities throughout the West, which themselves were a response to the growing demand from the Forest Service, the Soil Conservation Service, and other government agencies for technically trained professionals in range management. The formation of the Society for Range Management, in 1948, was further evidence of both trends, responding to a perceived need for greater professional autonomy from forestry and agronomy, whose schools and journals had hitherto housed most range training and scholarship. Taken together, these developments completed the formation, some fifty years after it began, of range science as an institutionalized field of research, education, and employment.

Stoddart, like Sampson, was a product of the University of Nebraska, and both textbooks explicitly described succession as the foundation of the science and management of rangelands. Both also identified livestock—the type and number of animals and the season of grazing—as the primary variable of concern to the range manager. Stoddart and Smith devoted an entire chapter to the subject, insisting that, "Correct numbers are of paramount importance to the perpetuation of the range, to the well-being of the livestock, and to the economic stability of the operator" (Stoddart and Smith 1943, 243). Neither book propounded fixed carrying capacities or static stocking rates, focusing instead on utilization rates while acknowledging that adjusting stocking to match variable forage production was difficult for ranchers as a practical matter. "In years of good forage, it is difficult to purchase livestock because of the demand from other producers. In poor years, the flood of animals on the market makes selling prices low. An attempt to regulate numbers according to range production by a buying and selling program is hazardous" (Stoddart and Smith 1943, 248). Finally, both texts treated shrub and brush problems—including mesquite in the Southwest, sagebrush in the Great Basin, and chaparal in California—as separate matters to be addressed through mechanical or chemical treatments or, in very limited circumstances, by careful use of fire. That these problems potentially violated the tenets of plant succession in relation to range management went entirely unremarked.

The grip of Clementsianism on range science was tight and persistent for reasons that were as much institutional as scientific. According to Tobey (1981, 152–53), the overall dominance of the Clementsian school peaked in the 1940s, persisting afterward only courtesy of the rise of range science: "Following World War II, grassland ecology was split into two professional interests. The long-term Clementsian synthesis of theory and practice was sundered, with practice split off, and range managers took most of the manpower out of grassland ecology. . . . The bulk of the specialty had shifted to technological concerns." Linda Joyce (1993, 135) describes it as "ironic that range scientists and managers were solidifying their use of climax while community ecologists were debating the very existence of community types and climax" in the late 1940s and early 1950s. Neil West (2003, 503) puts the matter more pointedly: "The Clemenstian model of plant succession was so dominant in the U.S. up to about 1950 that it was very difficult to publish anything contrary in either basic or applied journals. People in applied fields, such as range management, kept the Clementsian model alive much longer." This was not only because the discipline was "inwardly looking" and "backwardly focused," as Tobey, Joyce, and West all note, but because the scientists lacked an alternative theory that was adequate to *both* scientific *and* administrative-political demands. Practically speaking, experiments could only be done at smaller, shorter scales, and if the results could not be extrapolated, they would lack both scientific credibility and administrative utility. The sociospatial organization of the range itself—into pastures that were fixed, fenced, leased, and capitalized—meant that range science could scarcely exist without the assumption of linearity across scales of space and time. Succession in range science was less a theory than a policy.

FOUR

Fixing Stocking Rates

Monitoring and the Politics of Measurement

The problems that early range scientists encountered in the Southwest were particularly difficult to reconcile with successional ecological theory, but for practical purposes of management and administration, they were common to rangelands throughout the West. When Gifford Pinchot had resolved both to allow and to regulate livestock grazing on the national forests, he probably hadn't realized that he was committing the Forest Service to a task that would defy its best efforts for the better part of a century: the measurement of rangeland forage production in relation to livestock grazing. Throughout the western United States, the science of rangelands was preoccupied from its earliest days with measuring carrying capacities, utilization rates, and range conditions, all with an eye to determining proper stocking rates. At root this was an administrative problem, not a scientific one: some way of assessing *all* national forest rangelands was needed to render them visible and interpretable from local forest districts all the way to headquarters in Washington, DC, where officials sought to formulate uniform policies for administration and management. As we have seen, the need to fix stocking rates at static levels did not follow from Clementsian theory, nor did it align with the observations and recommendations of early range scientists in the Southwest. A 1937 bulletin put the problem bluntly: "Stocking southwestern ranges on the basis of the worst drought years is uneconomic and impracticable. Stocking on the basis of the best years is a suicidal policy" (Talbot 1937, 36). Nonetheless, static carrying capacities and stocking rates were deemed necessary within the larger institutional, political, and economic context surrounding public lands grazing.

In principle, all that was needed was accurate measurements, but the scale and diversity of the lands in question made this a daunting task both scientifically and logistically. What exactly should be measured, and how?

How should rangelands be classified and mapped? What were the appropriate units of forage production and consumption? How many men should comprise a survey crew, and what training did they need? How much land should they be expected to cover in a day? How should they interpret and report the data? How much would all of this work cost, and to what benefit? In the decades leading up to the Depression, the Forest Service wrestled with these questions as it sought to complete even a basic inventory of its rangelands; later, it and other federal agencies struggled—together at first, then in more or less divergent ways—to devise methods of monitoring rangelands to detect change over time, a related and in many ways still more difficult challenge (West 2003). Throughout, researchers and administrators faced perennial trade-offs between accuracy and cost.

Measuring and monitoring rangelands were never merely technical challenges, moreover. From the outset, reducing herds was seen as necessary to protect national forest rangelands, especially after large increases during World War I: more than 600,000 additional cattle and nearly 900,000 additional sheep in 1918 compared to 1914 (Jardine 1919, 40). But ranchers routinely resisted cuts, and their allies in Congress were powerful. The Forest Service needed accurate, rigorous, and defensible data, and it expected scientists to provide them. As historian William Rowley (1985, 111) observes, "Range science was the unspoken but necessary source of authority for aggressive range-management policies." James Jardine recognized this from the moment he was elevated to inspector of grazing for the Forest Service in 1910, and over the ensuing decades range scientists and administrators struggled to meet the often-conflicting demands of Congress, ranchers, and the larger public. What they reported publicly often differed from what they said internally in memos and conferences, not only because collecting data was labor-intensive and expensive but because it was nearly impossible to know how robust any of the available or proposed methods really were. Extrapolating from the scale of plots or transects to the scale of entire allotments was fraught with problems, as was extrapolating from experimental results to entire range types, forests, or regions. Indeed, the more carefully the scientists studied the matter, the more elusive their quarry appeared.

Range Reconnaissance and the Forage Acre

Beginning in 1912, Jardine dispatched crews out to the national forests, starting with those in the Southwest, to conduct what he called "range reconnaissance." Five years later, he reported that more than 12 million acres had been assessed, including more than 2 million acres in 1916 alone (Jardine 1916,

1; Jardine 1917, 1–2). The program was interrupted during World War I, resumed after 1919, and by 1925 some 23 million acres had been surveyed (Chapline 1925, 3). But this was still only about one-fifth of the total area of national forest rangelands; Jardine's 1919 annual report admitted, "At the rate reconnaissance work is now being done it would take from 50 to 60 years to cover all Forests by intensive reconnaissance" (Jardine 1919, 26). More rapid, "extensive" methods were adopted to expedite the process, and Jardine and Chapline repeatedly used their annual reports to plead for resources to pay for more crews.

The research conducted on experimental ranges was not as helpful in this effort as Forest Service officials had originally hoped. If carrying capacity varied widely from year to year on the Santa Rita and the Jornada, how much more variation could be expected across millions of acres of rangelands that differed in countless ways from those locations? Chapline stated as early as 1921 that carrying capacity could not be extrapolated from one range to another: "The problem of determining grazing capacity for the 25,000 individual ranges covering the widest possible variation, is perhaps the biggest one of grazing management. While considerable work has been done, its solution is a matter of years. Perhaps the biggest development, however, has been the realization that grazing capacity is a matter of study of the *individual range*, and the problem will not be completely solved until every one of our individual ranges is studied individually" (US Forest Service 1922, 130–31, emphasis in original).

Jardine and Chapline devoted as much funding as they could to range reconnaissance, and they refined both the methods of examination and the logistics for supporting crews to make them as economical as possible. Not only was progress slow relative to the overall extent of national forest rangelands, however, but by 1919 Jardine also realized that assessments would have to be repeated "at least once in three years with a view to adjusting grazing capacity on individual range units" (Jardine 1919, 17). At the District 3 Grazing Studies Conference in 1921, range reconnaissance was the first topic on the agenda and took up 40 percent of the conference's time, judging from the minutes (US Forest Service 1922). Conference participants included Chapline, the range staff from District 3—which comprised Arizona and New Mexico—and from both the Santa Rita and the Jornada Range Reserves, and one grazing official each from Districts 2 and 4. Detailed discussions took place regarding the size, composition, and training of reconnaissance crews; the definition of density, surface cover, and palatability; proper mapping techniques; expected rates of progress; and costs, which ran about one half cent per acre during the 1921 season.

Range reconnaissance could not achieve its administrative objectives unless the work of all crews conformed to a standardized protocol of methods and measurements that could be replicated in both space and time. Jardine's initial solution was a unit of measurement he called the "forage acre," defined as follows: "gross square acres surface supporting herbaceous and shrub vegetation × density (portion of the ground covered by foliage) × percentage of usable palatable vegetation ÷ one square acre = forage acres." The forage acre attempted to integrate the amount of forage produced, its accessibility and relative palatability to livestock, and the proper amount that could be grazed without damage to the plants. It postulated an imaginary ideal: an acre of land that was perfectly accessible to livestock (i.e., not too steep or too far from water) and completely covered with forage plants that were both maximally nutritious and capable of withstanding grazing. Although Jardine appears to have conceived this ideal as a pure abstraction—it resembled a tame pasture or perhaps the prairies of the Great Plains, but there is no indication that he had either one in mind—it anticipated the vision of perfect homogeneity in forage and grazing that would henceforth guide Forest Service range management on the ground (Fuhlendorf and Engle 2001). Starting from the imaginary ideal, range reconnaissance crews would make downward adjustments for lower densities of plants, sites that were less accessible, and plants that were less nutritious or less resilient (meaning that less of their biomass could be safely grazed). Jardine hoped that forage acres, expressed as decimals, could be determined for each type of range, just as foresters had determined ratios for converting diameter-breast-height to timber yield for different tree species. In theory, this would allow actual acres to be easily converted to a quantity of forage available for livestock, which could then be used to establish the number of animals that should be grazed there. Based on early measurements, Jardine estimated that one forage acre would support one sheep for 100 days, and that a cow required about 0.8 forage acre per month (Jardine and Anderson 1919, 28–29).

The forage acre proved difficult both to explain and to implement. In a footnote to *Range Management on the National Forests*, Jardine admitted that "This method of arriving at the grazing capacity of range is not in general use and therefore is not emphasized in the text. It is used, however, by grazing experts who have been trained in range reconnaissance surveys and estimates, and for this reason the forage-acre requirements of range stock are given" (Jardine and Anderson 1919, 27). Even the experts had a hard time making sense of the concept, however. In an undated memo, apparently written to another Forest Service range official, Jardine attempted to explain:

> It may be possible when we secure figures on carrying capacity by types to adopt some unit in place of the present forage acre unit which will have uniform value for all types and all localities. I doubt this very much, however, and I am sure it is not possible at our present state of knowledge regarding carrying capacity. . . . The only objection I have to it at present time is that it does not represent any unit area in the field. We find that in some cases ten actual acres are necessary to make one forage acre and in other cases within a mile the forage on two actual acres or even one acre of meadow may equal a forage acre. This is the fault of nature however.

That the forage acre was not a unit of area was counterintuitive, to be sure. Why call it an acre at all? But there were also deeper problems that could not be solved by changing the term or modifying the formula. For one thing, how livestock were managed could affect where and how much forage they could consume, as Jardine knew well from his experience with sheepherders in Oregon and elsewhere; he himself qualified his estimates as applying "to conditions of average management as regards distribution and control of cattle and herding of sheep" (Jardine and Anderson 1919, 30). For another, which plants livestock consumed was a matter not only of intrinsic palatability, fixed by the biochemistry of each species, but of what other plants were available in the vicinity; as Griffiths (1904, 25) had noted in his first report on the Santa Rita reserve, "it is exceedingly difficult to decide which species are and which are not forage plants. It often happens that nearly all plants that grow are eaten. What is grazed depends largely upon what is available for stock to eat within walking distance to water." A given herd's herbivory can also be affected by the animals' prior experiences and degree of habituation to that particular locale (Provenza et al. 2003). In other words, even if a typology of rangelands could be settled (which itself was a major question), available forage was not reducible simply to the density and composition of the plants in a given range type.

These problems were not really Jardine's fault, although calling them "the fault of nature" was rather disingenuous on his part. The diversity and variability of rangelands did pose enormous and perhaps insuperable obstacles to the quest for fixed and standardized measurements of carrying capacity across US rangelands, and the many other relevant variables made the challenge that Jardine faced extremely complicated. The fault lay not with the rangelands, however, but with the specific and entirely human constructs that made such measurements appear necessary in the first place. The problem was embedded in the concept of carrying capacity itself, understood as a static, quantifiable measurement of actual and/or potential conditions

(Sayre 2008). At root, the forage acre suffered from the problem of induction: Jardine asked range researchers and reconnaissance crews to measure real plants and real livestock, shaped by innumerable contextual circumstances, but the resulting ratios would then be used deductively as ideals to be applied by Forest Service administrators to other places and times—*X* animals on *Y* acres of range type *Z*—without regard to particular circumstances. It was an aspirational concept, as Jardine's memo candidly admitted, which he hoped scientific inquiry would someday perfect. That aspiration stemmed not from science, however, but from the need for "uniformity and a general approach" at the national scale of the Forest Service (Jardine and Anderson 1919, 2). An individual sheepherder or rancher would have no need for a carrying capacity abstracted from the particular circumstances that so strongly determined the growth and welfare of his or her animals. But for purposes of bureaucratic administration and the capitalization of rangelands, carrying capacities *had* to exist. The forage acre concept did not last long—it disappeared from publications and internal reports alike by the 1940s—but the underlying ambition that it expressed, and its attendant problems, would persist for the rest of the twentieth century.

The Taylor Grazing Act and *The Western Range*

As resources for range-related government programs increased, first under the 1928 McSweeney-McNary Act and then under Depression-era programs such as the Civilian Conservation Corps, the problem of monitoring and measurement spread from the national forests to all the federally owned rangelands of the West. It was of increasing importance for two major reasons. First, the question of the public domain—the lands not claimed for homesteading, transferred to the states, or set aside as Indian reservations, national parks, wildlife refuges or national forests—was coming to a head again, ultimately resulting in the Taylor Grazing Act of 1934. Some 128 million acres of these lands were grazed by livestock as open range without federal regulation of any kind, and they were both the least productive areas—those with better soils and more water having been disposed to private owners or reserved in the national forests—and by most accounts the most degraded due to uncontrolled grazing. That some system of regulation involving fences and grazing leases would be imposed on these lands was by this time accepted even by the livestock associations that had historically resisted such reforms. What was still uncertain was whether the lands would also be transferred from the Department of the Interior to the USDA, as had happened with the forest reserves upon creation of the Forest Service in 1905. The USDA had long argued for such a transfer, and it did so again following passage of the Taylor Act. On the other side were those who advocated the opposite: Interior Secretary Harold Ickes, supported for a time by President Franklin Roosevelt, wanted to change the name of his department to the Department of Conservation and absorb the national forests back from the USDA (Merrill 2002).

In this context, rivalries and contests between agencies, as well as with ranchers and livestock associations, prompted exaggerated claims of scientific knowledge about rangelands and how they should be managed. To make its case, the USDA assembled data from throughout the West into Senate Document 199, *The Western Range*, a 620-page "letter from the Secretary of Agriculture" to the Senate Committee on Agriculture and Forestry. The details were legion, but the basic argument was simple: national forest rangelands had stabilized and improved since 1905 under Forest Service management, while all other rangelands had not. As evidence, *The Western Range* presented quantitative claims about the 728 million acres of western rangelands. "The existing range area has been depleted no less than 52 percent from its virgin condition, using depletion in the sense of reduction in grazing capacity for domestic livestock. Practically this means that a range once capable of supporting 22.5 million animal units can now carry only 10.8 million" (Secretary of Agriculture 1936, 3). Three-quarters or more of the rangelands outside the national forests had declined in the preceding thirty years, compared to just 5 percent of national forest rangelands. Forest Service rangelands were still judged to be 30 percent depleted relative to "virgin" conditions, but the depletion of other rangelands was much greater: from 49 percent on state and county lands to 67 percent on the public domain. This apparent precision masked the fact that the underlying methods employed—unmentioned in the report—were little more than "a compilation of expert judgments" (West 2003, 502). Internally, Forest Service scientists and administrators struggled to define and measure range conditions for more than a decade afterward.

The Western Range ultimately failed to persuade Congress to transfer the lands in question to the USDA, but neither were the national forests transferred back to the Interior Department. Instead, the Department of the Interior created its own agency for administering grazing, the Division of Grazing (renamed the Grazing Service in 1939), which served this role until it was merged with the General Land Office in 1946 to create the Bureau of Land Management (BLM). If the debate gave the USDA and the Forest Service a strong interest in having—or at least appearing to have—authoritative and precise knowledge about rangeland conditions and grazing capacities, the outcome meant that there was now a second agency, in another department, that needed these kinds of information as well. Moreover, the Division of Grazing had even more rangelands to manage than the Forest Service did, and its lands were concentrated not in the wetter, higher elevations of the West but in the lower, arid and semiarid parts of the region. Under the Taylor Act, the sociospatial order that defined "the range" was extended to these lands.

A second reason for increased monitoring was that New Deal programs suddenly made unprecedented amounts of labor available to land management agencies of all kinds. The Civilian Conservation Corps in particular provided the Forest Service and the Division of Grazing with thousands of men whose labor could be deployed constructing fences, water sources, and erosion control projects and assisting with range assessments. (The crews also helped out with rodent and predator control, as we saw in chapter 1.) The Division of Grazing had so much land to survey, and so many Civilian Conservation Corps crews to keep busy, that it created an entire curriculum and textbook in range management for training purposes. With private, state, and federal lands forming a patchwork of jurisdictions and authorities, the need for coordination was acute, not only between the Forest Service and the Division of Grazing but with the newly created Soil Conservation Service (which was authorized to work on private lands), the Bureau of Indian Affairs, and numerous New Deal agencies.

Range Surveys: Publicly Professing Certainty

An Inter-Agency Range Survey Committee was convened in 1936 to address "the need for uniformity in range survey work leading toward the determination of carrying capacities." The committee agreed to use "range surveys" as the term for all such work, replacing "range reconnaissance," and to recognize two main types of surveys—intensive and extensive—based on existing Forest Service protocols. The following year, the committee issued a thirty-page pamphlet entitled *Instructions for Range Surveys*, which was ratified by the Western Range Survey Conference and signed by the heads of the Bureau of Indian Affairs, the Division of Grazing, the Resettlement Administration, the Soil Conservation Service, and the Forest Service. The purpose of the pamphlet was "to outline the present policy for the conduct of range surveys, and to standardize the methods used to the extent necessary to obtain the desired accuracy and uniformity in results." It began with a long list of factors that needed to be considered, from timber and water supplies to big game, recreation, and fire protection as well as grazing. "The closest integration and coordination of these uses are essential if serious conflicts are to be avoided. As the demands for the various uses increases conflicts develop, the settlement of which requires accurate information regarding all the factors involved. . . . This activity is essentially field laboratory work in range management" (Inter-Agency Range Survey Committee 1937, 2).

The instructions for extensive range surveys were basically those of the Forest Service's earlier range reconnaissance program, but with "forage acre" renamed and restricted to internal purposes only. Guided by the best available maps (at a scale of one or two inches per mile), crews were instructed to traverse rangelands on foot or horseback, following cadastral section lines laid down by the General Land Office's surveyors. Each section would be traversed on the section and half-section lines and mapped according to "forage types," each map unit being at least twenty acres under most conditions. "In passing through the type the examiner will mentally calculate and carry with him a moving average of plant density and composition" based on visual (or "ocular") observations, looking directly downward and estimating the portion of the soil surface covered by vegetation. Once the overall density was thus determined, the examiner would estimate the composition of the vegetation by major species and add up the cover values of each as a check on the overall estimate. Finally, "After the composition rating for each individual species has been recorded, that rating is multiplied by the accepted palatability rating for the species, and the sum of all the individual products yields the weighted average palatability of the type. This last figure multiplied by the estimated density yields the forage factor or palatable density of the type." The forage factor was to be "used in compilation of the data but should no longer be placed on the final map or used in grazing capacity summaries" (Inter-Agency Range Survey Committee 1937, 12), apparently because its meaning was so difficult to explain.

Intensive surveys differed from extensive ones by measuring a sample of plots with a modified quadrat rather than visually inspecting an entire map unit. "The square foot density method is a system of sampling vegetation by randomized and replicated plots" (Inter-Agency Range Survey Committee 1937, 13), each plot being a circle of radius 5.64 feet, enclosing an area of 100 square feet. Within each plot, the examiner would estimate—again visually, but aided by a one-foot-square frame—the density of each species of plant present, imagining the individual plants bunched together at the ideal of complete coverage envisioned in the forage acre. The figures for each species would be added up to compute density for the plot and divided from the total to determine composition. Areas of 10 to 20 acres would be sampled with three plots, those of 20 to 80 acres with five plots, and those from 80 to 640 acres with ten plots. Plot locations would be determined based on the overall size and distribution of different range types; a rock could be thrown to locate a plot "at random," or a line through a unit could be established and plots located at predetermined intervals in a strip or grid configuration. The resulting data were averaged across all plots in a map unit, then subjected to the same calculations of palatability and density as in the extensive method.

In theory, the intensive method was more accurate than the extensive one because it studied small areas more precisely, relying on principles of sampling to represent larger areas faithfully. Nonetheless, there was considerable room for judgment and potentially for error, and a follow-up study (Reid and Pickford 1944) later found evidence that the intensive methods actually produced more variability between examiners than the extensive methods did. The basic observational practice—looking directly downward and estimating the area covered by plants—was the same in both methods, and examiners were further instructed "to visualize the plant as it should appear at maximum growth in a normal climatic year" (Pickford 1940, 6)—that is, "the plants should be mentally reconstructed to compensate" for amounts grazed, amounts yet to grow by year's end, and abnormal rainfall (Inter-Agency Range Survey Committee 1937, 14). This was obviously a major methodological concession, but how else to account for the practical fact that surveys took place at various times of year, grazing was ongoing, and forage growth varied from one year to the next?

There was also the problem of deriving the conversion factor—Jardine's forage acre, or what the instructions called "the forage acre requirement." How much of the various plant species, combined in various proportions in a given type of range, could safely be grazed without compromising future forage production? The best the committee could recommend was to find fenced areas "that have every appearance of having been properly used for a period of years," were representative of the larger range, and for which surveys had been done. "At the close of the season the chief of party will make utilization and range condition studies of these ranges and will obtain the most accurate and detailed information possible on the rate of stocking and seasonal use that has obtained on such areas for the past several years" (Inter-Agency Range Survey Committee 1937, 18). In other words, a fenced, grazed area in apparently very good condition would supply the norm of proper stocking for a type of range; capacities for other areas of the same type could then be computed by adjusting for discrepancies in the density and composition of the forage as determined by surveys.[1] The instructions advised setting capacities slightly lower than the computed figures indicated and adjusting them in future years based on the results in a given site.

Ultimately, then, the determination of grazing capacities was a trial-and-error process, informed by data but strongly conditioned by subjective considerations both in measurements and in the selection of reference sites. This reality was obscured, however, on the very next page of the instructions, where the procedure was described in the seemingly fixed terms of an equation: "To compute the grazing capacity, multiply the surface acreage of a type by its forage factor, and divide the result by the proper forage acre requirement. The forage acre requirement may be in terms of sheep or cow months, or years. Grazing capacity is, therefore, expressed in sheep or cow months, or years, according to the forage acre requirement used" (Inter-Agency Range Survey Committee 1937, 19). The AUM, or animal-unit-month, became the standard unit for expressing grazing capacities from this point on; one AUM is the amount of forage that will support one adult cow, horse, or cow-calf pair or five sheep or goats for one month. But the forage acre requirement was no longer a biophysical constant for each range type—as in Jardine's original vision of the concept, modeled on conversion factors for timber species—but a variable, one that would henceforth disappear from the very documents that depended on it for their production:

> The forage acre has erroneously been accepted as a constant. Actually it is a variable. This is evident because of the continual need of applying different forage acre requirements to obtain grazing capacity in different localities or in the same locality with different methods of estimation. Consequently, the forage acre has been misleading to stockmen, to economists who have attempted to capitalize it, and to agencies who have attempted to correlate the grazing capacity on different ranges. In the future, forage acres will be omitted on all range maps and Graphic plans and grazing capacity in terms of animal months substituted therefor. (Inter-Agency Range Survey Committee 1937, 19)

"A Mathematically Precise Base Which Does Not Exist": Privately Admitting Uncertainty

The *Instructions for Range Surveys* were "the first and last time [that] uniformity in procedures was agreed upon" among federal land management agencies. "Every action since this period has produced divergences" (West 2003, 503). This may be due in part to the fact that the Inter-Agency Range Survey Committee was concerned only with assessing the current state of rangelands, not with the more difficult tasks of detecting and explaining changes in vegetation or range conditions over time. The committee's instructions said nothing about plant succession or climax communities, and they expressed a confidence that surveys, if properly conducted, could resolve the question of grazing capacity. This was important in an interagency context, as the Forest Service sought to assert its experience and expertise vis-à-vis the newer agencies. But in the relative privacy of a Range Research Seminar at the Great Basin Experiment Station in 1939, Forest Service scientists and administrators struck a markedly different tone. Succession occupied pride of place in the program segment on "Fundamental Concepts," motivated at least in part by a desire to get away from the subjective and inductive approach tacitly endorsed by the Inter-Agency Range Survey Committee. "Such a study of succession, of ecological relationships and changes, is fundamental for interpreting range problems and is necessary for developing proper methods of range management aimed at improving the range vegetation. Without such knowledge, range management can only go ahead in a blundering trial and error fashion without scientific basis" (US Forest Service 1939, 159).

Despite all the work that had already been done, the problem of determining carrying capacities seemed no closer to solution in 1939 than it had for David Griffiths in 1904. Indeed, in many ways it appeared more difficult than ever before. Everyone at the seminar agreed that quantitative measurements were needed, but the more they worked to refine their methods, the more sources of error they discovered. George Stewart, head of range research for the Intermountain Forest and Range Experiment Station, remarked that extensive and intensive surveys each had advantages and disadvantages: The former ensured that "a large part of the forage is given at least a passing glance" but yielded estimates of composition that "may be uncertain unless carefully checked; and the method may not reveal inconsistencies in the data that are due to personal error." Intensive survey methods permitted more statistical analysis but risked being unrepresentative of larger areas and "tends to become more and more intense" and therefore more expensive. Consistency of methods between individuals was valuable, Stewart pointed out, but it did not necessarily improve the accuracy of the measurements. "When it comes to accuracy, the answer is even more elusive, as it sometimes seems that the problem will defy solution until a means is found of obtaining the true value which will make it possible to say how closely this true value is approached by a given survey inventory." In other words, the methods could not be evaluated without knowing the "true value," but there was no way to determine the true value other than the methods themselves. For the time being, "Just how accurate range surveys really are is a difficult question to answer with precision. . . . In our present stage of scientific development in range there is no ready means of learning what the true mean really is. Therefore, it is extremely difficult to learn the exact magnitude of error involved, but it can be approximated at least roughly. . . . Our best indications are that the error in forage inventory is around 15 to 20 percent—and in some cases possibly a little more" (US Forest Service 1939, 166–68). Without greater accuracy, converting forage production to stocking rates was fraught with error: "The foundational information is the inventory. The conversion factor is the forage acre allowance. This factor must be in direct proportion to the densities and proper use levels used. Unless this is done locality by locality, the numbers derived may be in error. To derive this factor mathematically necessitates a mathematically precise base which does not exist" (US Forest Service 1939, 181).

Stewart and others at the seminar called for improved sampling techniques to reduce error. "Just what these sampling methods and measurements or units should be, is a question for which there is as yet no satisfactory solution," said Gordon Merrick, associate forest ecologist from the Northern Rocky Mountain Forest and Range Experiment Station (US Forest Service 1939, 161). Stewart pointed out that the number of plots was more important than their size: "the natural variability in surveys is so great in any case that there is little wisdom in using much time on small variables when the need for replication and randomization is so important." When comparing two sites—ungrazed and grazed, for example—"size and shape of plot are of little consequence, if the plot is large enough to give a fair cross-section and small enough to be recorded rapidly. Replication of plots per site (10 to 50 usually used) and replication of comparisons are the really important points" (US Forest Service 1939, 170–72). The trade-offs were simultaneously scientific and budgetary: precision and accuracy were possible at small scales, "for example, a 160-acre pasture. But when the processes required are transferred to a range of 250,000 acres, the samples, of necessity, must be spread so thin that observation supplants measurement again" (US Forest Service 1939, 179).

The magnitude of the problem was subject to debate. Assistant Regional Forester Earl Sandvig, speaking from the perspective of administration, argued that extensive survey methods were sufficiently accurate in relation to the object of concern, namely grazing: "Even though ocular methods of estimating are relatively crude and are repeatedly subjected to attack, I think it can be safely said that our harvesting machine, in the form of cows and sheep, is even more crude when compared to the refinement we go to now in measuring the forage crop. . . . Keeping in mind the crudity of the machine that harvests the crop, there is little advantage in building a measuring machine to a much finer degree of precision" (US Forest Service 1939, 179, 182). Sandvig, among others, opined that range surveys, however flawed, were "an extremely useful administrative tool in the accomplishment of proper range management" (US Forest Service 1939, 182). Stewart insisted that, "even with the errors, surveys are still the most reliable means of adjusting the rate of stocking to grazing capacity" (US Forest Service 1939, 168).

Even if all the technical problems of measurement were resolved, however, the natural variability of forage production would still pose a challenge to effective administration of rangelands. Whatever the average, or "base," grazing capacity of a range might be, actual conditions over time would expose those estimates to disbelief or even ridicule. In disputes with stockmen, Sandvig complained, "The focal point of attack becomes the grazing capacity derived from the survey, if it differs materially from the existing estimate, because it is a vulnerable and important element. An excellent job of range survey can become all mired down over differences of opinion regarding grazing capacity." Grazing capacities were "based on average forage production and not upon the production existing during the year of examination. Such a level of grazing capacity necessarily will result in under-or overuse during other than average years." Given the politics of measurement, even Forest Service officials might come to doubt the data if they failed to understand these limitations. "The ideal, of course, is to determine the size of the forage crop a year ahead and use this for setting the number to be grazed. While ideal, such a method is impossible at present. The average crop base has appeared to be the best alternate method. But unless the method used is fully understood by the administrator, he is prone to compare the determined average capacity with the current year's production and, if it varies, to declare it wrong" (US Forest Service 1939, 180).

The committee report on grazing capacity at the close of the seminar attempted to square this circle by distinguishing between long-term change and short-term variability. "As ranges are restored from a depleted condition as a result of good management, grazing capacity will increase. On the other hand, overgrazing causes grazing capacity to decline. Variations in forage production from year to year due to weather fluctuations do not change grazing capacity" (US Forest Service 1939, 398). But this left unanswered how the two kinds of change—each occurring on its own temporal scale—were related to each other, and it implied that weather played no role in restoring degraded rangelands. The core conclusion of the committee was, in effect, a recitation of the bewildering array of factors that had to be considered:

> Research on what constitutes proper use of key forage species with relation to their resistance to grazing and to soil stability, correlated with more definite knowledge on livestock preference of plants growing in association with key species, and preferably combined with estimates of production expressed in pounds of dry herbage per acre, will more nearly allow true estimates of grazing capacity in range surveys than is possible at present. . . . Range surveys should estimate the value of vegetation present on the range, after which grazing capacity should be estimated through proper deductions or "utilization cuts" for slope, soil erosion, game, recreation, timber utilization, and so on. No mathematical formulas are known at present to properly evaluate these factors as they affect grazing capacity. (US Forest Service 1939, 399)

In 1940, one of the seminar participants published an article in the British journal *Herbage Reviews* on "Range Survey Methods in Western United States." G. D. Pickford was senior forest ecologist at the Pacific Northwest Forest and Range Experiment Station in Portland, Oregon, and his article echoed both the 1937 *Instructions for Range Surveys* and several of the views expressed at the seminar in 1939. He acknowledged that fluctuating rainfall complicated efforts to assess grazing capacity; that the extensive and intensive survey methods were subject to error in various ways; and that the ideal underlying the forage acre "does not exist in Nature" (Pickford 1940, 8). Notwithstanding these shortcomings, however, Pickford concluded that range survey data provided "a most systematic and impersonal appraisal of range values, range problems and range conditions" and that the lack of such data had hitherto "seriously handicapped" the improvement of range management. He reported that approximately 37 percent of the total range area in the western United States had been surveyed as of January 1, 1939, including 138 million acres of private land, 22 million acres on Indian Reservations, 51 million acres of the national forests, and 57 million acres on lands administered by the Division of Grazing. "The recent, marked increase in coverage of western grazing lands by range surveys cannot but add to the effectiveness of future resource plans and management" (Pickford 1940, 11). By Pickford's figures, though, 63 percent of western rangelands still remained terra incognita to government administrators, including 37 million acres in the national forests and 85 million acres of Division of Grazing lands.[2] The methods employed in the surveys that had been done, moreover, were potentially less accurate than anyone in the relevant agencies cared to admit publicly.

The Rangeland Conflict and Quantifying Range Condition and Trend

For political as well as administrative purposes, having some knowledge, however flawed, was better than having none. World War II caused the Civilian Conservation Corps to be disbanded, and during the war both surveys and research were curtailed. But the postwar years saw the politics of western rangelands flare up again. Western and national livestock associations launched a concerted campaign to devolve Grazing Service rangelands from federal control to the states, counties, or private ownership. Eastern congressmen objected to the below-market rates charged for grazing leases, viewing the deficit between expenditures and receipts as an unfair subsidy to public lands ranchers. In 1946, Congress cut the agency's budget by half, prompting the Department of the Interior to merge it with the General Land Office to create the Bureau of Land Management (BLM). Things escalated further when *Harper's* editor Bernard DeVoto launched a series of articles and trenchant editorials denouncing ranchers as greedy despoilers of public lands. DeVoto's writings helped galvanize a national constituency to resist public lands ranchers in the name of the nation's patrimony and public trust, and the modern "rangeland conflict" was born. Officials from the Forest Service and the BLM found themselves caught in the middle of a string of public hearings and congressional committee meetings that continued through most of the decade.

Historians have correctly noted that the pitched debates of the 1940s hinged on competing notions of property rights and on grazing fees—how much ranchers paid per animal-unit-month to graze their livestock on public lands (Merrill 2002, Peffer 1951, Foss 1960a). The Taylor Act was internally contradictory on the property rights issue, and courts subsequently ruled that, although grazing leases were privileges, not rights, ranchers could nonetheless mortgage and sell their leases as though they were, in effect, their own property. Grazing Service fees were even lower than Forest Service fees, both were below market rates for private rangelands, and the two agencies struggled mightily to determine the proper value of their forage relative to market rates for private lands (Rowley 1985, Merrill 2002). But if property rights and grazing fees were conspicuous issues of public policy, readily subject to legislative and judicial reckoning, they also obscured—for partisans and historians alike—the actual source of the dispute, which range surveys were supposed to resolve: the determination of permitted stocking rates on grazing allotments. This was the ranchers' core economic concern because it determined both how much a lease was worth (as collateral or for sale) and how many AUMs a rancher could obtain by the payment of fees (at whatever rate). In other words, a fee increase raised the cost of doing business, but a reduction in the grazing capacity of an allotment devalued the ranch itself. According to ranchers, federal range officials were demanding cuts in stocking rates that were arbitrary and capricious.

The problem was the very one that the 1939 seminar participants had identified but not solved: the distinction between long-term change and short-term variability, which the idea of carrying capacity collapsed. When rains and forage were plentiful, ranchers saw good conditions and sought to increase their herds, whereas the agencies considered such fluctuations transitory and epiphenomenal. As Marion Clawson (1948, 267), the first director of the BLM, observed, ranchers and range scientists defined range condition in very different ways: "By range condition the latter means the relation between the present productive capacity of a range and its potential capacity; condition changes slowly as a range slowly deteriorates or is restored, while forage production may vary sharply and quickly as weather factors dictate. Upon this concept, it is unthinkable that range condition would fluctuate several points from month to month, or seasonally, as the rancher's variety of 'range condition' will be shown to do. The rancher obviously has something else in mind."

Although the Forest Service and the BLM ultimately retained regulatory control of federal rangelands, the pressure for accurate and defensible scientific knowledge about grazing capacities and range conditions grew ever more acute. It was in this context that two advances converged and appeared to provide the long-sought synthesis of successional theory with practical methods of range survey and measurement. The first was an article published in the recently founded *Journal of Range Management* in 1949 by E. J. Dyksterhuis, a former Forest Service range examiner then working for the Soil Conservation Service in the Great Plains of Nebraska, Oklahoma, and Texas. "Condition and Management of Range Land Based on Quantitative Ecology" presented a method of quantifying the successional stages described by Sampson thirty years earlier in Forest Service Bulletin 791; it soon supplanted that bulletin to become the most important article in range science for the next forty years.

Dyksterhuis divided range plants into three categories: those that increased in frequency under livestock grazing, those that decreased, and those

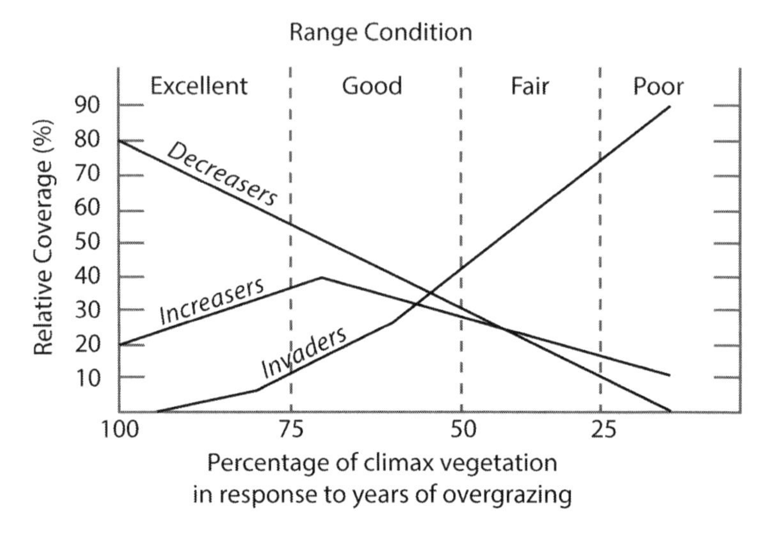

4.1. A diagram of range condition classes as a function of departure from climax proportions of three classes of plants: those that decrease under grazing pressure, those that increase, and those that invade as the climax species disappear from a site. Redrawn from Dyksterhuis (1949) by Darin Jensen.

that invaded highly degraded areas, replacing both increasers and decreasers. "All range plants belong to one of these groups," he wrote, and data showed "that upward and downward movements along the curve of range improvement and deterioration were continuous series of changes in the relative proportions of decreasers, increasers, and invaders" (Dyksterhuis 1949, 108–9).[3] With these categories in mind, range condition could be quantified relative to the climax plant community for a given site: each species' contribution to current plant cover would be credited up to its share in the climax community, and the sum would represent a percentage of climax. Condition classes could then be defined as excellent, good, fair, and poor by quartiles: 75 to 100 percent of climax being excellent, 50 to 75 percent good, and so on (figure 4.1).

Separate from Dyksterhuis's work, the Forest Service launched a three-year project in 1948 known as the "condition and trend study," whose "primary purpose was to find for administrative use a method or methods which would be reasonably simple, practical, accurate, and technically sound, and yield concrete measurements as well as sound observational evidence on the trend or rate of change in range condition" (Parker and Harris 1959, 55). The study involved more than fifty researchers, but the result came to be known as the "Parker 3-Step" after Kenneth W. Parker, the Forest Service range scientist who led the effort. Step 1 was a quantitative measurement of plant density and composition using 100-foot transects, with a three-quarter-inch loop placed on the ground at each foot along a metal tape. The loop was examined for bare soil, litter, or a plant, and the nearest perennial species to each loop was also tallied. Step 2 used the resulting data and observations of plant vigor, shrubs, and erosion to assign a numerical score that could be used to classify a site into condition types (excellent, good, fair, and poor). Step 3 involved taking photographs at transect locations for visual documentation. By comparing the assessments from fixed plots over time, trend could be determined.

The bulk of the study concerned the assessment and refinement of Step 1, including trials of different methods of measuring vegetation to reduce examiner variation and improve efficiency without sacrificing accuracy. Tests determined that the three-quarter-inch loop was better than larger loops (which introduced undue variation in readings) or discrete points (which missed single-stemmed plants) and that the results correlated very closely (with a coefficient of 0.9719) with the line-intercept density method (Canfield 1941), which was "generally considered a reliable index of the actual percentage of ground occupied by vegetation" (Parker and Harris 1959, 58) but was more time-consuming to conduct (figure 4.2). How precisely

4.2. Measuring vegetation with the line-intercept method, Santa Rita Experimental Range, 1943. Source: Santa Rita Experimental Range archives.

transects could be relocated and replicated over time, and with what effects on the data collected, were also studied carefully, as was the loop method's sensitivity to changes in plant density and composition at sites scattered across seven western states. Notably, even though studies suggested that twenty to twenty-five clusters of randomly located transects were needed to detect a 20 percent change in vegetation with 95 percent confidence, "this intensity of sampling was prohibitive from an administrative standpoint," and a minimum of two clusters per range type and range condition was recommended instead (Parker and Harris 1959, 65).

Dyksterhuis's article came out while the condition and trend study was ongoing, and Parker integrated the increasers-decreasers-invaders classification into Step 2 of the method. "Range condition classes are in reality successional stages of plant communities described in terms to make them readily available for use by range administrators and stockmen" (Parker 1954, 14). In 1951, the three-step method was presented and endorsed at a meeting of Forest Service administrators and researchers and at a separate gathering of livestock industry representatives, state university researchers, and other state and federal agencies; in 1956, the Forest Service adopted the Parker 3-Step for range analysis across the national forests (Parker and Harris 1959, 64–65). The BLM, however, retained the monitoring methods of the Inter-Agency Range Survey Committee, thereby reintroducing the very thing the committee had been convened to prevent: a disparity between the two agencies' data about rangelands (National Research Council 1994).

At midcentury, it appeared that the problem of rangeland monitoring and measurement had finally been solved, at least as a scientific matter (Joyce 1993). Excellent, good, fair, and poor became the official rangeland condition classes for both research and administration, determined by reference to a fixed climax for a given type of range. Underpinning the categories was a successional framework that rested on three key assumptions. First, the ideal condition for virtually all rangelands was the climax, understood as a single fixed state for any given site, dominated in most cases by perennial grasses. Second, the climax was equated with the conditions that prevailed prior to Anglo-American settlement and the arrival of livestock; "we assume the optimum development of vegetation as being usually synonymous to the original vegetation" (Parker 1954, 16). Third, the key variable determining range condition was livestock grazing. Other factors—such as rainfall or drought—were assumed to be transitory and secondary, capable of producing "noise" in forage production but not of determining the true signal of range condition, which according to successional theory was fundamentally about plant species composition.

Ironically enough, it is impossible to know exactly how many livestock actually grazed on western US rangelands for most of the twentieth century, even though the allotment files of the Forest Service and the BLM are replete with such figures. Many allotments were large, rugged, and remote, and cattle could virtually disappear for months on end. With fences built and predators controlled, many ranchers saw little need for regular tending by cowboys, and there was no need for the cattle to bunch together for self-defense. Individuals or small groups of animals might wander off, and in extreme cases herds could become effectively feral. Some ranchers probably didn't know themselves how many animals they had on the range until roundup each year, and even if they did know, they had no reason to give the agencies an accurate accounting. If they had more than the permitted numbers for the lease, they could face penalties and risked losing the lease. If they had fewer, they might have their permit reduced—or at least many of them feared as much—under "use it or lose it" guidelines intended to ensure that forage was not wasted. It was preferable for ranchers *and* administrators that the paperwork faithfully reflect the official numbers prescribed in a lease, every year, so that eyebrows were not raised and the fees paid and received were those expected and budgeted for by both parties. And this is what one finds, far more often than not, in the files: constant stocking rates year after year, changing only when a ranch—with its lease—was sold to a new owner, and a cut—typically of 10 percent—was imposed as a condition of reissue to the new lessee. As a result, it is also impossible to know how much actual stocking rates may have varied in response to the vagaries of rainfall and forage production. Several range researchers noted in passing that ranchers knew perfectly well to reduce their herds during drought (Renner 1948), and the massive livestock die-offs of the late nineteenth century did not recur during the droughts of the 1920s, 1930s, and 1950s. Presumably, most ranchers did not want to see their animals die of thirst or starvation, and they may have adjusted their herd sizes accordingly. It is unlikely, however, that many herds varied nearly as much as rainfall and forage.

The convergence of successional theory, range survey methods, and the political and economic interests of agencies and ranchers was overdetermined, but it was neither preordained nor inevitable. Indeed, even at mid-century there could be heard a small chorus of range scientists from *outside* the Forest Service who publicly criticized range survey methods precisely because of the static stocking rates that those methods were used to establish. Laurence Stoddart, for example, noted that methods of measuring forage were "not sufficiently accurate for use in research, nor have they been satisfactory to or accepted by stockmen." Interannual forage variability was

300 to 1,000 percent, Stoddart observed, but other factors supervened on federal rangelands:

> Federal range managers are forced to grant long-term permits for specific livestock numbers. Specific numbers are demanded by the livestock growers to stabilize the industry, and only through occasional and minor adjustment in the duration of the grazing season can even slight adjustment to forage production be made. . . . This is, in many ways, a wasteful philosophy of grazing capacity because, if forage is to be adequate in years of low production, then there must be excessive amounts of ungrazed forage in years of high production. We have no reason to believe that overuse in one year can be compensated by underuse in another year; hence, stocking that would result in correct use in the average year does not appear to be the answer. (Stoddart 1953, 1368)

Scientists in the Soil Conservation Service were still more scathing in their critiques of static stocking rates. J. L. Lantow and E. L. Flory, writing in the agency's official journal *Soil Conservation* in 1940, argued that "sound economics and the objectives of soil conservation cannot be reached with the continuance of a set number of livestock or a prescribed plan of use" (Lantow and Flory 1940, 140–41). Annual precipitation could vary by an order of magnitude in the Southwest, they noted, and "Extreme as climatic fluctuations have been shown to be, forage production varies still more" (138). They called for "yearly and seasonal changes of management" and "adjustments in livestock numbers . . . to make the range use fit the fluctuations in production" (137). The forage available at the end of the growing season should be measured and assessed against the number of animals dependent on it for the coming year, they wrote. If the forage was insufficient, "adjustments should be made immediately, either by securing more feed or by reducing the number of livestock in accordance with the available feed" (141). In a thinly disguised jab at the Forest Service, Lantow and Flory wrote: "Sustained grazing capacity is set by some agencies at various percentages below the average forage production of the range as determined by range surveys. But this average forage production is very difficult to establish with any degree of certainty, and, if correct, might not be the best figure to use as a base for any given year" (142).

Whatever the actual stocking rates on Western rangelands were, over the second half of the twentieth century, the problems and limitations of the Parker 3-Step and successional theory became evident on the ground. The institutionalized premise that stocking rates were the key variable driving range conditions turned monitoring into a kind of cat-and-mouse contest with ranchers. Highly variable rainfall meant that the visible effects of stable stocking could vary from harmless to destructive, and distinguishing short-term variability from long-term change remained elusive. Battles with ranchers over stocking rates did not abate, and Forest Service files swelled with letters from ranchers demanding to graze more animals during wet years when forage was abundant, alternating with memos from range staff complaining during dry years that the herds were excessive and the grass was gone. The Forest Service responded by tasking scientists to refine monitoring methods and protocols to improve their statistical rigor (Sayre, Biber, and Marchesi 2013). But it eventually proved impossible to resolve the core political and economic issue—how many AUMs federal agencies would permit—simply by recourse to the apparent scientific authority of ever-more-precise measurements. As early as 1960, historian Phillip Foss (1960b, 64) wrote: "The process involved in weighting and evaluating these various preferential 'rights' [i.e., setting carrying capacities] . . . are tremendously complicated and time-consuming for the grazing district staff, and are subject to varying interpretations which may in turn lead to disputes and litigation."

In 1961, eight years after the Jornada Experimental Range was transferred out of the Forest Service, scientists there published the conclusion that "Sustained grazing capacity does not exist on these semi-desert ranges" (Paulsen and Ares 1961, 83). One of those scientists, Fred Ares, later wrote: "In retrospect, of all these early attempts to apply proper and effective range management, it can be said, 'One thing thou lackest!' That is the failure to place more emphasis and stress on timely stocking adjustments. The old concept of a fixed carrying capacity pretty much prevailed and was clung to and followed to the bitter end, regardless of current conditions" (Ares 1974, 22). By the late 1980s, the Forest Service had largely abandoned allotment monitoring (Government Accounting Office 1991). More "rigorous" data, it turned out, were not necessarily more effective in persuading ranchers to change their management, and Forest Service employees' time might be better spent building relationships with permittees than measuring forage (Sayre, Biber, and Marchesi 2013). This left the agency empty-handed, however, when faced with widespread litigation by environmental groups in the 1990s. In a two-part study of the history and methods of US rangeland monitoring, Neil West (2003, 530) summarized the core problem this way: "Many have assumed that information could be easily scaled either upward or downward. It should now be clear . . . that these assumptions are usually invalid." In 1994, a committee of the National Academy of Sciences concluded that the condition of the nation's rangelands could not be known due to "the lack of current, comprehensive, and statistically representative data obtained in the field" (National Research Council 1994, 3). Methods of rangeland monitoring remain an active topic of research and debate to this day (Havstad and Herrick 2003; Pellant et al. 2005).

FIVE

To Manage or Manipulate

Natural versus Artificial Improvement of Depleted Rangelands

On January 18, 1945, Jerome O. Eddy of the Diamond and Ahalf Ranch near Prescott, Arizona, wrote a letter to his three congressmen, Senators Carl Hayden and Ernest McFarland and Representative Richard Harless. "As you know I am a stockman," he began. "I am very much interested in producing as much as possible on my ranch. You know how the Forest [Service] officials act. As soon as they think that a range is over grazed, all they know is order a cut of cattle numbers." Alluding to "an experiment of juniper eradication on my ranch" conducted by scientists from the Santa Rita Experimental Range, Eddy proposed that an appropriation be made for the Forest Service "to improve ranges from reseeding and increase the carrying capacity instead of cutting down herds to satisfy the ranges."

Eddy was a prominent Arizonan—a member of the state's 1944 Republican Party delegation and reportedly an heir to the Quaker Oats fortune—and the response to his letter was swift. On January 22, McFarland forwarded the letter to the head of Forest Service research, E. I. Kotok, and Harless sent it directly to the secretary of agriculture. The next day, Hayden—a member of the powerful Senate Appropriations Committee[1]—forwarded it to the chief of the Forest Service, L. F. Watts, attached to a letter of his own in which he wrote:

> Mr. Eddy is of the opinion that additional money should be appropriated by the Congress for the use of the range reserve program of the Forest Service, particularly with regard to the work being done by the Santa Rita Experimental Station near Tucson, Arizona. Mr. Eddy feels that the best possible service that can be performed in connection with range experimental work would be to undertake the reseeding of western range lands to increase their carrying capacity, thereby eliminating a part of the necessity of cutting down on grazing permits because of poor range condition.

> I am inclined to the opinion that this is a sound idea and I shall be much obliged to you for any comments or advice that you feel will be helpful to me in preparing a responsive reply to Mr. Eddy.

On January 24, W. R. Chapline attached Hayden's letter to a memo marked "confidential" and "airmail" to the director of the Southwestern Forest and Range Experiment Station, alerting him that "We will prepare a reply indicating the opportunity for increasing grazing capacity through range reseeding and the need for additional research to develop effective procedures for the important range types in need of reseeding." One week later, Forest Service Chief Watts replied to Hayden. "The Forest Service fully agrees with you and Mr. Eddy that a more adequate program of reseeding investigations on western ranges would pay real dividends. Reseeding will be necessary on some 80 million acres of deteriorated range of the West, to restore the forage cover, protect the soil, and permit the development and maintenance of productive and efficient range-livestock operations." Reseeding research had previously been limited to the intermountain and northern Rocky Mountain regions, he asserted, resulting in "selection of desirable adapted species as well as development of economical and effective methods of planting in several of the more widespread range types." Some three million acres had been reseeded, mostly in those two regions, increasing "grazing capacity by from 2 to 10 times. Thus, land that does not produce enough forage to justify the taxes has been converted to productive range." But in the Southwest, Watts wrote, progress had been more limited. Work by the Civilian Conservation Corps had been "valuable in pointing out some of the problems peculiar to the region and in indicating the possibilities for success," but with the termination of the CCC, "the work had to be prematurely ended. Facilities have not been available for even a reasonably small program of reseeding in the Southwest." He went on to explain: "The results of investigations from other regions cannot be widely applied in the Southwest because of highly variable soils, scant and unevenly distributed precipitation, and extremes in temperature. We have done enough to know that these factors combine to make especially difficult reseeding problems that will require detailed investigation. . . . Such studies, if initiated promptly, should provide stockmen and land-management agencies with information on where, what, how, and when to plant, so that real progress in reseeding could be made when manpower and machinery are again available." Kotok wrote a similar letter to Senator McFarland on February 6, and Assistant Secretary of Agriculture Charles Brannan adapted Kotok's letter to reply to Representative Harless on March 15. Reseeding had worked elsewhere, they all reported, and it would work in the Southwest if only the research could be done. With funds in the offing, the Forest Service's southwestern range research program leaped into action, and by June a series of reports titled *Problem Analysis—Range Reseeding* had been drafted. In the years that followed, grasses would be collected and sown, machinery would be developed and tested, and shrubs and trees would be removed or poisoned, not only in the Southwest but across many of the West's rangelands, supported by the augmented funds that flowed, it seems, from Eddy's well-placed letter.

In their replies to the congressmen, however, the bureaucrats in the Forest Service had been rather selective, if not altogether misleading, in their presentation of what had already been done and what could be expected. The agency actually knew a great deal about reseeding southwestern rangelands, and most of it was discouraging. Research had, in fact, been ongoing almost continuously since the Division of Agrostology sent its first investigators to the region at the turn of the century. Hundreds of native and imported grass species had already been seeded or transplanted in trials using myriad techniques, and the results had usually been similar: Reseeding worked with only a few species and only in the most favorable circumstances, and even apparent successes were often reversed during subsequent droughts. Summarizing reseeding research in his 1916 annual report, inspector of grazing James Jardine (1916, 2) had written: "The work as a whole confirms previous opinions that the seeding of range to cultivated species as a practical means of increasing the forage will be limited to a very small percentage of the total range lands unless species are secured which are better adapted than those at present available to range conditions." In 1931, the USDA had published a circular on *Artificial Reseeding of Western Mountain Range Lands*, which concluded that no species could be expected to succeed without at least seventeen inches of average annual precipitation and at least six inches during the main growing season—effectively ruling out all but the highest elevation sites in the Southwest. "In general, artificial reseeding tests with native and cultivated plants on the range in southern Arizona have thus far been distinct failures," the authors wrote (Forsling and Dayton 1931, 11, 30). In 1937, Chapline (1937, 8, 11) had written: "there is little chance for improving the bulk of native range lands in their present condition by the introduction of the common cultivated forage plants," although he was cautiously optimistic about native grasses. "Thousands of dollars have been wasted by stockmen," he complained, "in unsound efforts at artificial reseeding." As recently as 1940, a short "research note" published by the Southwestern Forest and Range Experiment Station repeated what researchers had stressed for decades: "It is more practical and economical to secure recovery on depleted areas through natural means than by artificial seeding" (Parker and McGinnies 1940).

The two broad strategies evident in these passages characterized range science from the earliest work of the Division of Agrostology. "Natural" improvement was about *managing livestock* such that natural processes would fix range problems: desirable plants would expand, undesirable ones would decline, erosion scars would heal, water tables would rise, and the full suite of rangeland values—forage, timber, watersheds, recreation, and wildlife—would all flourish. "Artificial" improvement, by contrast, was a more aggressive and agronomic approach that involved actively *manipulating rangelands*: cultivating or fertilizing the soil, modifying the terrain to slow erosion and capture water, and identifying cultivars—native and nonnative, domesticated and wild—that could be selectively improved in greenhouses and field trials, produced in large quantities, and then sown or transplanted to increase forage on depleted rangelands.

The underlying contradiction between the two strategies was never openly acknowledged. Whereas the first relied on the Clementsian assumption that nature reliably restores itself if relieved from excessive disturbance, the second replaced it with a conviction that other, *more* intensive disturbances were needed to remake nature to suit human needs and demands. Cultivating and seeding large areas of the range were *de facto* admissions that secondary succession toward climax conditions was not happening (or at least not quickly enough) and that nature could not solve range problems unaided. Federal agencies were more concerned with practical than theoretical problems, however, and they sidestepped the contradiction by dividing degraded rangelands into two categories: those that could improve "naturally" through management, and those that had lost so much of their forage plants and topsoil that more expensive manipulations were necessary. As Watts wrote to Hayden, the area in the latter category was estimated in the 1940s as 80 million acres (cf. Pearse 1943; Chapline 1948), or a little more than 10 percent of the nation's rangelands.[2]

Where and how to draw the line between the two strategies, and between the corresponding categories of land, took a marked shift after World War II. Up until that point, artificial reseeding had generally failed, especially in drier areas, and even where it had worked, its gains had proven temporary without improvements in livestock management. Natural reseeding had therefore been given priority over manipulating rangelands. At midcentury, however, even as successional theory consolidated its central place in academic range science, widespread, conspicuous failures of natural improvement tipped federal range research toward artificial, capital-intensive interventions. The failures were greatest in precisely those areas where Clementsian theory foundered: the drier and more variable rangelands of the Southwest and the Great Basin, where fencing and stocking rates based on carrying capacity estimates had not caused succession (back) toward climax; instead, "noxious plants" such as mesquite (*Prosopis velutina* and *P. glandulosa*) in the Southwest and big sagebrush (*Artemisia tridentata*) in the Great Basin appeared to have thickened and expanded at the expense of former grasslands. Despite fifty years of mostly discouraging results, federal agencies doubled down on artificial improvement, taking advantage of post–World War II prosperity, cultivars from overseas, new chemicals and equipment, and political support for technological solutions. Heavy machinery was developed for clearing, cultivating, and seeding rangelands, and methods of aerial seed and chemical application were perfected. Millions of pounds of grass seed—mostly of nonnative species—were produced, collected, distributed, and planted.

Efforts in the Great Basin met with early apparent success, and they seem to have revived federal agencies' determination to find similar solutions for degraded southwestern rangelands. Big sagebrush did not resprout from the root crown, so it could be removed with relatively simple mechanical means

5.1. Removing sagebrush with the Olson sagebrush rail near Sublette, Idaho, 1943. Source: National Archives Forest Service photograph 428776. By Joseph F. Pechanec.

5.2. A test plot of imported grasses planted in the sagebrush steppe of the Great Basin. Source: National Archives Forest Service photograph 448345.

such as a steel rail dragged behind a tractor (figure 5.1) or with fire (Renner 1951),or, before much longer, with chemicals (Hyder 1954). And crested wheatgrass (*Agropyron cristatum*)—a perennial forage species imported from Asia by the USDA around 1900 and widely planted on abandoned croplands in the northern Great Plains in the 1930s—was successfully seeded on former sagebrush rangelands as early as 1932 (Hull and Klomp 1966; figure 5.2).

Sagebrush control and crested wheatgrass seeding expanded dramatically through the 1950s and 1960s, motivated by both the promise of greater forage and the need to control two other nonnative plants that were invading degraded Great Basin rangelands: *Halogeton glomeratus*, a member of the amaranth family that is poisonous to livestock, and cheatgrass (*Bromus tectorum*), a prolific annual grass that dramatically increased the risk of fires. The history of crested wheatgrass (and cheatgrass) in the Great Basin has been well documented (Dillman 1946; Knapp 1996; Pellant 1996; Young and Clements 2009), but the case of southwestern brush control and artificial reseeding is less well known (Roundy and Biedenbender 1995).

As Eddy's letter suggests, the appeal of artificial improvement was heightened by the impasse between ranchers and agencies regarding stocking rates and range conditions. The failures of the newly ascendant, successional paradigm of range management would not force changes in the paradigm until decades later (chapter 7); in the meantime, those failures were converted into opportunities for researchers and agencies alike. If ranchers and agency officials were perennially at odds over whether and how natural reseeding took place—unable to disentangle the relative effects of variable rainfall and contested stocking rates—the two groups could find common cause in the aspiration to remake rangelands on the model of agriculture—provided, that is, that the expense could be justified to politicians, bureaucrats, and ranchers. As with fencing earlier, success or failure was ultimately defined by economic rather than ecological criteria, and the greatest problem was scaling up from the plots and pastures where research was conducted to the tens of millions of acres deemed in need of transformation. Because technological manipulations could potentially become cheaper as use expanded or as new discoveries were made, there was always the possibility of future success, regardless of past failures. And as earlier with rodents, predators, and fire, the precondition of conservation soon became extermination, this time applied to shrubs.

Early Attempts at Artificial Reseeding

Efforts to improve degraded rangelands by sowing and cultivating grasses began not with scientists but with settler ranchers and farmers. In the late 1890s, when H. L. Bentley conducted the Division of Agrostology's first investigation of range and forage plants, he reported that stockmen in central Texas had already tried "to grow alfalfa and other forage plants not native to the country. With irrigation alfalfa has done well, but it has been found impracticable to grow it successfully and profitably under other conditions in this section" (Bentley 1898, 17). Johnson grass (*Sorghum halepense*) grew better but was a scourge to farmers whose fields it invaded; the "objections to its extensive propagation . . . are so well known that they need not be enumerated in this connection" (Bentley 1898, 17–18). Australian saltbush (*Atriplex semibaccata*), clovers, and Kentucky bluegrass (*Poa pratensis*) had also been tried, without success; milo maize and sorghums required replanting every year, "and this involves no little labor and expense. What stockmen need," Bentley insisted, "are hay meadows of *native* grasses that have shown in past years all the best qualities of the best hay grasses elsewhere, and that do not require any experimental work to determine their adaptability and general value. Such an investment would be a paying one" (Bentley 1898, 18, emphasis in original).

Bentley's emphasis on native grasses both reflected and resisted the agronomic tendencies of the Texas stockmen and the USDA alike. The standard model of agricultural improvement at the time involved the use of domesticated plants, almost all of them brought from Europe or elsewhere. The famous grass botanist George Vasey, in his *Report of an Investigation of the Grasses of the Arid Districts of Kansas, Nebraska, and Colorado* for the USDA's Division of Botany in 1886, expressed the widespread view that cultivation was invariably more productive than unaided nature:

> Man has learned to select those plants, grains, and grasses which are best adapted to his wants, and to grow them to the exclusion of others. This is the essence of agriculture. Nature shows her willingness even here to respond to the ameliorating influences of cultivation. . . . There is every reason to expect that even the gramma [*sic*] grass may be made to double its yield by cultivation. But there is a considerable number of grasses native to this district which are much more thrifty and productive than the gramma and buffalo, and if they were selected and sown upon the properly prepared land there can be no doubt that a great improvement in the grass production would be effected. Indeed we should extend our inquiry to foreign grasses cultivated in similar situations. (Vasey 1886, 10)

The federal government had been collecting germplasm from around the world since before the USDA was created in 1862; one of the department's core activities was testing plants in greenhouses and experimental gardens and distributing promising cultivars to farmers in places where climatic and soil conditions resembled those where they originated (Kloppenburg 1988). The majority of the Division of Agrostology's work nationwide was along these lines: cataloging grass and forage species from across North America and around the world and cooperating with state experiment stations to test them at locations all over the country. The superiority of exotic plants was widely considered to be obvious: they matched the expectations of Anglo-American settlers (and of the markets in which they sought to sell their crops), and in the eastern United States they had fared well, aided by the suite of other imported organisms that accompanied them (Crosby 1986). They were also assimilated to prevailing ideologies of racial and civilizational progress: Old World crops, like Old World peoples, were considered naturally superior to their New World counterparts.

By contrast, western rangelands presented the possibility that native plants might actually be better for livestock production than imported ones. But it was still assumed they would have to be domesticated and cultivated. In the Division of Agrostology's first circular, Jared Smith (1895, 2)—who had earned a graduate degree from the University of Nebraska in 1892—observed that "Nearly all of our cultivated grasses and clovers are of foreign origin," and he lamented the fact that so few native species had yet been studied for possible cultivation. Circular number 4, published the following year, asserted that "The fact that cultivation improves the more desirable native grasses has been demonstrated by nearly every experiment station in the West and by a great many private parties as well" (Williams 1896, 2). In the same vein, Bentley (1898, 18) asked, "Why should stockmen look to foreign countries or even to other sections of Texas for grass seeds and hay? . . . Here, with so many varieties needing no experimental work to determine their value, we seem bent on destroying them as speedily as possible. While others in less-favored sections [i.e., Europe] are developing one new variety we are systematically destroying a dozen quite as good. Let us take care of what we have and develop them." In support of these claims, Bentley described ranchers who had had success with simple techniques of scattering native grass seed before the rains or onto melting spring snow, breaking up bare soil with a harrow, or plowing furrows that captured wind-blown seeds. *"Native grasses are by far the best for home use; they are suited to the climate and the climate is suited to them"* (Bentley 1898, 21, emphasis in original).

Over a period of three years at the turn of the century, Bentley conducted the first government experiments in range improvement, under the direction of Jared Smith. A stockman by the name of C. W. Merchant lent them the use of 640 acres of his private, degraded lands near Abilene, Texas, and the people of Abilene donated fencing materials and two water tanks (Smith 1898, 2). The land was fenced into ten pastures of ten to eighty acres to implement various treatments: removing livestock, grazing at different portions of the year, or alternating grazing and rest; some pastures were harrowed or disked, and the smallest was reserved for cultivating both native and introduced plants. After one year, Bentley estimated a 25 percent increase in carrying capacity on the pasture that had been disked and grazed compared to surrounding ranges, but he also found that alternating grazing and rest "secures the same result at much less expense" (Smith 1899, 22). Attempts to establish forage by seeding were unsuccessful the first two years but promising in the third year, when the rains were better. Asserting "that two comparatively unfavorable seasons are very likely to be followed by one at least that is specially favorable"—he called this "a general fact" but had no further evidence for it—Bentley concluded: "Taking an average of three years, it is practically certain that excellent results can be depended on if correct methods, based on correct ideas, are pursued" (Bentley 1902, 27).

Native grasses outperformed imported ones in both transplantation and sowing trials; central Texas stockmen and farmers, Bentley wrote, should "place less reliance on the oftentimes extravagant claims made by interested dealers in seeds, etc., as to imported grass and forage plants, and devote more attention and time to those native to their respective sections" (Bentley 1902, 30).

The relative contributions of natural and artificial reseeding were impossible to determine, however, because the results were the aggregate and confounded outcome of both. Advised by three local stockmen with long experience in the area, Bentley continuously adjusted stocking rates in response to changing forage conditions. "It was never possible during the three years that the station work was continued to carry out very strictly the definite plans for treating the station pastures as originally laid down" (Bentley 1902, 22). Indeed, he inadvertently performed what may have been the first documented experiment in rotational grazing in the sense since popularized by Allan Savory (chapter 7):

> At no time . . . were the stock permitted to graze all of the pastures at the same time. The effort was made to give each pasture periods of rest in regular succession. That meant the doubling up of the stock in the pastures in which they were being held; hence it happened that during considerable periods of time the different pastures were carrying quite double as many animals as the average recommended by the inspectors. As one result of this systematic resting, the grasses in each pasture were, to a greater or less extent, permitted to mature seeds, which, falling to the ground, increased the number of grass roots and in that way added materially to the capacity of the range for supporting stock. (Bentley 1902, 34)

The more intensive, artificial treatments, including cultivating the soil by disking and "sowing the seeds of hardy native and improved grasses" (Bentley 1902, 26), occurred on the same pastures as the grazing and rest periods, making it impossible to disentangle their respective effects. Bentley thus advised ranchers to do all three—disk, sow, and rotate their herds: "Every stockman should see to it that instead of one or two large pastures he should have a number of small ones, some of which can be resting while the others are doing, perhaps, double duty" (Bentley 1902, 35). This recommendation had remarkably little influence on subsequent research, however, probably due to the high cost of fencing rangelands into so many pastures, especially on larger scales and in more rugged terrain.

The Division of Agrostology's circulars reported optimistically on plant introductions in many parts of the country where rainfall was greater than in central Texas, and Bentley, too, expressed hope that several imported plants could be cultivated in "tame" pastures. But early attempts at artificial reseeding in drier areas of the Southwest were not only unsuccessful but inconsistent, suggesting that factors beyond human control were of overriding significance. David Griffiths cultivated forty different species of forage grasses on plots within the tract near Tucson that preceded creation of the Santa Rita Range Reserve, but all failed during the 1898–1904 drought. He later concluded that rainfall patterns were the key variable: some plants responded to winter rains, others to summer rains, and two successive good summers were required for perennial grasses to establish successfully. Rather than attempt to reseed perennials on areas dominated by less valuable forage plants such as annual grasses and weeds, he advised replacing cattle with goats or sheep and avoiding overgrazing until the perennials were able to reestablish on their own (Griffiths 1907, 22). Similarly, Jardine's report on his 1914 examination of the Jornada Range Reserve concluded that seeding and plant introductions should be kept to a small scale, "and should be considered of minor importance until the problems of management and carrying capacity are well under way." Will Barnes echoed Jardine in a memo the following year. Natural reseeding, he wrote, "should be made the major project at both the Jornada and Santa Rita Reserves, and should be pushed as aggressively as possible."

The pattern persisted well into the 1930s: artificial reseeding of rangelands showed promise in higher, wetter parts of the West and in unusually wet years in lower, drier regions, especially where nonnative grass species were well adapted to local conditions. Jardine's annual report for 1916 expressed cautious optimism from artificial reseeding experiments in Utah, but not for the Southwest. Any efforts larger than at "an experimental scale" were simply too expensive given the uncertainty of success (Jardine 1916, 2). This remained the case three years later: "In range management we are dealing with large areas, usually of low grazing capacity, and on such lands the extent to which improvements can be made is limited both by natural conditions and by expenditures proportionate to the value of the forage" (Jardine and Anderson 1919, 43). Jardine's (1919, 3, 2) annual report concluded that seeding experiments at the Jornada with both cultivated and native species "have proved failures" and predicted that artificial reseeding "will be limited largely to a very small, but important, percentage of the total range lands where growth conditions are exceptionally favorable." In 1933,

Chapline (1933, 9) addressed the World Grain Conference and summarized some 600 reseeding experiments: "The results indicate that mountain range areas of Utah and the central and northern Rocky Mountains, several million acres in aggregate, having superior growing conditions but where native plants have been practically destroyed, can be reseeded profitably to several cultivated and native forage plants." A 1938 study on the Santa Rita Experimental Range used various mulches to retain soil moisture in sites seeded to ten native grasses; germination on the mulched sites was four to twenty times higher than on control sites (Glendening 1942), but mulching, like cultivation, was too expensive for large-scale application.

If there was any room for optimism, it came mostly from outside the Forest Service. In the 1920s, the New Mexico College of Agriculture and Mechanic Arts (now New Mexico State University in Las Cruces) had had some success with small-scale reseeding at sites scattered across the state, although rodent control, cultivation, and erosion control were necessary to assist establishment (Wilson 1931). By the early 1940s, scientists there believed that "the limit of rainfall where artificial reseeding will be successful may be as low as 9 inches" (Bridges 1941, 3). The Depression and the Taylor Grazing Act drew additional federal agencies into range management, as we saw in chapter 4, and by 1942 the Soil Conservation Service (SCS) and the Bureau of Plant Industry had identified three imported grass species that showed promise for "regrassing" southwestern rangelands to aid in erosion control (Flory and Marshall 1942). Both the state and federal scientists had also developed techniques and equipment for harvesting, cleaning, and sowing grass seeds and for modifying soil and water conditions to encourage establishment (figure 5.3; cf. Hoover 1939). Despite this progress, researchers emphasized that artificial revegetation was costly and should be used only in severely degraded sites where natural improvement was impossible, and that proper range management was prerequisite to success in all cases: *"It is useless to attempt range restoration as long as the period of grazing and degree of use which depleted the range in the first place are continued"* (Flory and Marshall 1942, 17–18, emphasis in original). With these provisos, however, they asserted that "There are many sites in the Southwest where artificial revegetation may be used as an aid in the control of erosion with reasonable assurance of success" (Flory and Marshall 1942, 8).

The most successful of all the grasses that had been tested by this time for southwestern rangelands was Lehmann lovegrass (*Eragrostis lehmanniana*), seeds of which had been obtained from South Africa by the Bureau of Plant Industry in 1932. After initial plantings showed promise at the Boyce

5.3. The "one-man seed gatherer" for harvesting seeds from perennial grasses to be used in artificial reseeding projects, Utah, 1938. Source: National Archives Forest Service photograph 374163. By Alvin C. Hull, Jr.

Thompson Southwestern Arboretum in Superior, Arizona, the SCS's Nursery Division undertook systematic studies of Lehmann and two other African lovegrass species[3] in 1935, and ten years later Franklin J. Crider published the results as a USDA circular. There were some 250 lovegrass species known from throughout the world, Crider noted, but they "have not been recognized as of much agricultural importance in the United States. Judged solely from the standpoint of forage value . . . this view may be justified. However, when evaluated upon the broader basis of usefulness for soil and moisture conservation combined with crop and livestock production values, some of the more recently tested lovegrasses appear to have a definite place in the agriculture of many parts of the country. Particularly do they merit consideration during the present War emergency period because of qualities which enable them to produce quick, effective results" (Crider 1945, 2).

Crider reported that all three of the grasses "come up quickly from seed, grow fast, mature rapidly and early, form deep, dense root systems, set good seed crops, self-seed naturally, [and] are very drought resistant" (85). Their forage value was deemed comparable to "the best native and tame grasses," at least when green and tender (87), and they withstood grazing well. Lehmann lovegrass stood out from the other two "in that it proved to be the most easily established and rapid-spreading of any grass tested for revegetation purposes under the very severe, semi-desert range conditions of the Southwest. . . . In addition, it was the only grass to become fully reestablished by self-seeding upon the return of normal rainfall after being killed by prolonged drought" (86).

Crider did not consider that any of the African lovegrasses might displace native perennial species in the United States, but his circular hinted strongly at that possibility. He characterized all three as "belonging to the earlier or pioneering stages of grassland succession," based on observations that other species dominated the climax stage in South African settings—at least this was true in the moister, eastern portions of their ranges. On the other hand, however, "they become dominant in the climax stages of the semi-open grasslands of the drier West," and in disturbed sites, they "often become dominant and persist for a long time" (Crider 1945, 5–6). These words, apparently overlooked or considered unimportant at the time, would prove prescient.

Artificial Reseeding after World War II

In the decade following 1945, the Forest Service rapidly expanded its range reseeding research effort, and an Interagency Range Reseeding Committee was formed to coordinate efforts with the Soil Conservation Service and the recently renamed Bureau of Plant Industry, Soils, and Agricultural Engineering. Minutes from the committee's fourth annual meeting, in 1951, suggest that the pace and magnitude of expansion were a strain. Large areas were being seeded based on preliminary experimental results, for example, without waiting for research to determine the geographic limits or grazing sensitivity of different varieties. The seed supply was a major topic of discussion: each study required seeds of sufficient quantity and quality, and the species and varieties sought changed as new results came in. Ramping up production took time, and the SCS's Nursery Division struggled to meet the agencies' demands for the seeds considered best suited to the diverse settings and changing techniques of reseeding.[4] In 1955, nearly 2.5 million pounds of seed were requested, almost double the amount of just three years before; more than half of the total were wheatgrasses, including 727,000 pounds of crested wheatgrass alone.

In the context of suddenly increased resources, it was useful both to downplay the extent of previous research and to express optimism about future prospects. Kenneth Parker's 1945 draft *Problem Analysis—Range Reseeding in the Southwest* opened with the assertion that "the experimental work in the

Southwest to determine practical methods of reseeding and suitable species has been desultory and sporadic. Effective and sustained research work as yet largely remains to be done." But by 1949, the Southwest Forest and Range Experiment Station had produced *A Preliminary Guide for Range Reseeding in Arizona and New Mexico* (Reynolds, Lavin, and Springfield 1949), which described where, how, and with what species reseeding could be done. Various species of wheatgrasses were suitable in higher-elevation rangelands with average annual precipitation of fifteen inches or more, but in the semidesert grasslands, shrub control had to precede reseeding, which was recommended "only where average annual rainfall is above 12 inches." The full problem analysis, produced in 1948 and revised in 1950, claimed that research since 1945 had "demonstrated that range reseeding can be an important adjunct in restoring a much needed plant cover on certain deteriorated range lands." But it also cautioned: "The information now available with regard to reseeding applies only to the better sites in the areas of higher rainfall" (Glendening and Parker 1950, 2–3).

The challenge was huge: less than 3 percent of Arizona rangelands wasn't deteriorating, two-thirds was "severely deteriorated," and an estimated 2 to 15 million acres would need to be reseeded (Glendening and Parker 1950, 15–16). In hindsight, figures such as these must be treated with care, because they reflected the methods of range condition classification and measurement described in chapter 4, which are now known to have contained important flaws. At the time, however, they were considered authoritative. The assessment asserted that "range reseeding . . . can be counted upon to result in a rapid positive improvement of the range resource on at least the more critical areas" (Glendening and Parker 1950, 15). The benefits of reseeding included not only increased forage and livestock production but reduced flooding, erosion, and sediment flows into the growing number of dammed waterways in the region.

As had been the case since Vasey's and Bentley's time, the underlying model for reseeding research was cultivated agriculture. The studies began with intensively controlled, garden-like conditions and proceeded in stages toward more realistic range settings. The Forest Service's *Working Plan for Reseeding Research in the Southwest* called for a series of trials at thirty-five study sites established since 1945 across Arizona and New Mexico. The first step was plantings of individual species in three rows, twelve feet long and one foot apart, with each species separated from its neighbors by three-foot buffers. "Seedbed preparation will be intensive. All competing vegetation will be removed and a firm seedbed will be prepared by disking, harrowing, and cultipacking. Nurseries will be fenced against livestock and rabbits, weeded once a year, and plants otherwise given every opportunity to become established." The row plantings would determine "preliminary adaptability of a species to climate and soils as measured by forage production, seedling vigor, and capacity for reproduction" (Reynolds and Bridges 1950, 5–6). In the second step, species that showed promise would be planted on one-tenth- to one-quarter-acre "yield plots" in randomized blocks, again with careful seedbed preparation, with one block "left undisturbed to determine yield of remnant native vegetation" (Reynolds and Bridges 1950, 8). In step three, the species that succeeded in step two would be incorporated into mixtures, planted again, and yields measured for two growing seasons, with attention paid to the spacing and timing of seeding and the amount of seed used per acre. Only then would a mixture proceed to step four: grazing trials on forty-acre sites within national forest allotments to test for the degree of grazing, animal weight gain, and season of use.

Artificial reseeding soon became more about mechanical and chemical engineering than about ecology. Cultivated agriculture was already highly mechanized in the United States and rapidly becoming more so, and chemical fertilizers and pesticides proliferated after the war. But rangelands posed challenges for which existing agricultural technologies were not well suited: rugged terrain, rocky or shallow soils, and, above all, a massively increased spatial scale. Seedbed preparation was seen as necessary, for example, but it was also expensive and potentially damaging. "In most cases broadcasting the seed without soil treatment is wasted effort. It is very important to get the seed into the soil. However, most range plants should never be planted more than ½ to 1 inch deep. . . . Plowing is generally too expensive and, furthermore, is apt to result in serious erosion from either wind or water" (Parker and McGinnies 1940, 2; cf. Renner 1951, 545). The working plan envisioned experimenting with existing equipment of all kinds, and the Forest Service's Equipment Development Laboratory was enlisted to invent new ones. "The ideal equipment for preparing a seedbed on range land should be capable of destroying a maximum of competing vegetation with a minimum of draft, and should result in a firm, trash-covered seedbed" (Reynolds and Bridges 1950, 10). The need for "effective cheap methods which can be used to cover large tracts of country" was "critical," and "the possibilities of the airplane as a rapid means of dispersing seed in otherwise inaccessible areas or during short periods of favorable climatic conditions should be fully investigated" (Glendening and Parker 1950, 3).[5] Methods of seeding would include "the conventional methods of seeding used by farmers, such as the grain drill, broadcasting with a hand seeder and covering with a disk, spike-tooth harrow, or brush drag. . . . In addition, tests will be made with the special grass seeding drill, the Blackland Planter, cultipacker seeder, and mechanized broadcasters such as those used for distributing insect bait or poison" (Reynolds and Bridges 1950, 11). And although previous research suggested that fertilizers were "economically unfeasible on low-value range lands," the working plan argued that "the possibility . . . that fertilizers may mean the difference between success and failure in reseeding should not be overlooked." Tests with nitrogen, phosphorus, and potassium would be conducted at different application rates to determine "whether or not fertilizers assist in establishing reseeded species" (Reynolds and Bridges 1950, 20).

By 1957, researchers from the SCS, the Rocky Mountain Forest and Range Experiment Station, and the University of Arizona recommended reseeding of southwestern desert grasslands receiving at least eleven inches of annual rainfall. In a bulletin aimed at ranchers, they wrote: "Many southern Arizona desert-grassland ranges are producing much less forage than they could and should. The productivity of many of these range lands can be restored practically and economically by artificial reseeding with adapted forage plants. In some instances, reseeding has quadrupled forage yield" (Anderson et al. 1957, 7). The grass they recommended above all others was Lehmann lovegrass, planted by a "cultipacker seeder" to loosen the soil and cover the seeds, and accompanied by rodent and insect control.

The Mesquite Problem

Even if the seeds, equipment, and procedures for reseeding grasses could be perfected by research, by this time another problem precluded improvement of many southwestern rangelands. As we saw in chapter 3, rather than inducing successional recovery of climax forage plants, controlled stocking rates in fenced allotments had instead been accompanied by continuing increases in the extent and density of shrubs, brush or woody plants (figure 5.4). Study sites on the Santa Rita and Jornada Experimental Ranges demonstrated that even complete livestock exclusion did not necessarily result in grassland recovery—indeed, shrub invasion could be even greater on ungrazed than grazed sites (Glendening 1952). The 1957 bulletin cautioned that, "where mesquites exceed fifteen to twenty-five trees per acre they should be eradicated or at least thinned" prior to reseeding.

Before artificial reseeding could succeed, something would have to be done about the shrubs. In the decade before he developed his famous

5.4. Photographs taken from the same place on the Santa Rita Experimental Range in 1903 (top) and 1941 (bottom), showing mesquite encroachment. Source: National Archives, Forest Service photos 402888 (by David Griffiths, March 10, 1903) and 425038 (by Matt Culley, 1941).

5.5. W. R. Chapline, Kenneth W. Parker, and Arthur Upson discussing shrub control on the Santa Rita Experimental Range, 1938. Source: Santa Rita Experimental Range archives.

three-step method of range monitoring, Kenneth Parker authored internal planning reports on the control and eradication of snakeweed, burroweed, and mesquite for the Southwestern Forest and Range Experiment Station (figure 5.5). In the *Journal of Range Management,* he described noxious plants as "an ever-increasing threat to the welfare and permanence of the western livestock industry" (Parker 1949, 128). Within the southwestern region of Arizona, New Mexico, and Texas, the "semidesert grassland and shrub" zone—some 84 million acres in area and concentrated between 3,000 and 5,000 feet in elevation—was emblematic of the challenge: not only had it lost a greater proportion of its forage than other zones, but it had proven more difficult to improve. Once dominated by open stands of highly nutritious grasses, it was now particularly degraded, with 90 percent considered to be in a downward trend (Glendening and Parker 1950, 32). The most promising cultivars were all nonnative lovegrasses from Africa: Boer, Lehmann, and Wilman. But reseeding could not succeed, even with these species, without first addressing "the invasion of mesquite, cholla, and burroweed.

5.6. Mesquite clearing with circular saw, 1947.
Source: Santa Rita Experimental Range archives.

Practical reseeding measures on such sites must await the development of an economical method for destroying these noxious plants" (Glendening and Parker 1950, 53)

From the 1940s well into the 1960s, the problem of mesquite dominated southwestern federal range research, especially at the Santa Rita Experimental Range (Sayre 2003). By 1948, researchers had tried grubbing, girdling, or cutting down mesquites and poisoning them with kerosene, diesel fuel, or arsenic (figure 5.6). A 1952 USDA circular estimated that mesquite was the dominant plant on more than 70 million acres of southwestern rangelands, and that at least half of that area had been invaded in the past century; it called mesquite invasion "one of the most serious and perplexing problems in southeastern Arizona" (Parker and Martin 1952, 1). Quadrat measurements made on and adjacent to the Santa Rita showed that both crown cover and abundance of mesquites had more than doubled from 1932 to 1949, while grass cover had declined more than 90 percent. The causes of invasion were multiple and complex, but comparison of grazed and ungrazed quadrats revealed that "grazing by livestock had no effect on the rate of mesquite increase after the invasion had begun" (Parker and Martin 1952, 11). Even mesquites just two years old had deep roots that enabled them to outcompete perennial grasses during drought periods; beyond about 7 percent mesquite crown cover, grass cover began to decline, dropping to almost zero where mesquite reached 30 percent.

Shrub control resembled and extended the logic of extermination that had already been applied to predators and rodents; indeed, removing rats and rabbits was seen as necessary, both to keep them from spreading the seeds of shrubs and to ensure they did not eat the grass seeds used in reseeding projects. Killing mesquites wasn't easy, but experiments indicated that if it were done, the grasses would quickly rebound. In the late 1930s, sodium arsenite showed promise in tests on ranches in southern Arizona; white arsenic, acid arsenic, sulfuric-acid sprays, petroleum oils, sodium thiocyanate, and ammonium sulfide were also tested on mesquite and other shrubs (Streets and Stanley 1938). A study begun in 1940 used sodium arsenite to kill all mesquites in one-acre plots at four locations on the Santa Rita; in some plots, burroweed plants were also removed by grubbing. Within three years, the cover and forage yield of perennial grasses had doubled on the treated plots compared to untreated control areas. A follow-up study, begun in 1945, reduced mesquites to twenty-five, sixteen, nine, and zero plants per acre on two-acre plots; native grass production over the following five years increased 200 to 400 percent where mesquites were completely killed; the fewer mesquites, the more the grasses increased (figure 5.7). These results did not obtain on sites where grasses were absent, however: if perennial grass cover was less than about 10 percent, reseeding with Lehmann lovegrass was recommended. Moreover, because new mesquites would continue to germinate, treatments would have to be repeated, becoming "a permanent feature of ranch operations in large areas of the Southwest. . . . Invasions, if not attacked in the early stages, may later require many dollars for control" (Parker and Martin 1952, 64).

As promising as these results appeared, they were achieved at scales of only a few acres, and scaling them up remained dauntingly expensive. To be effective, sodium arsenite had to be applied by hand to individual plants, and each plant had to be girdled or a basin dug out around the base before the poison was applied (figure 5.8). These steps made killing mesquite on large scales very expensive (prison labor was used for some of the Santa Rita experiments), and to make matters worse, sodium arsenite and its fumes were extremely toxic to humans and animals—so much so that the wood of killed trees could not be safely burned. Diesel oil and kerosene were safer and less labor-intensive, but they were less effective and still required application to every plant individually. Parker and Martin (1952, 61) estimated the cost at eight to fourteen cents per plant in 1947, and they presented

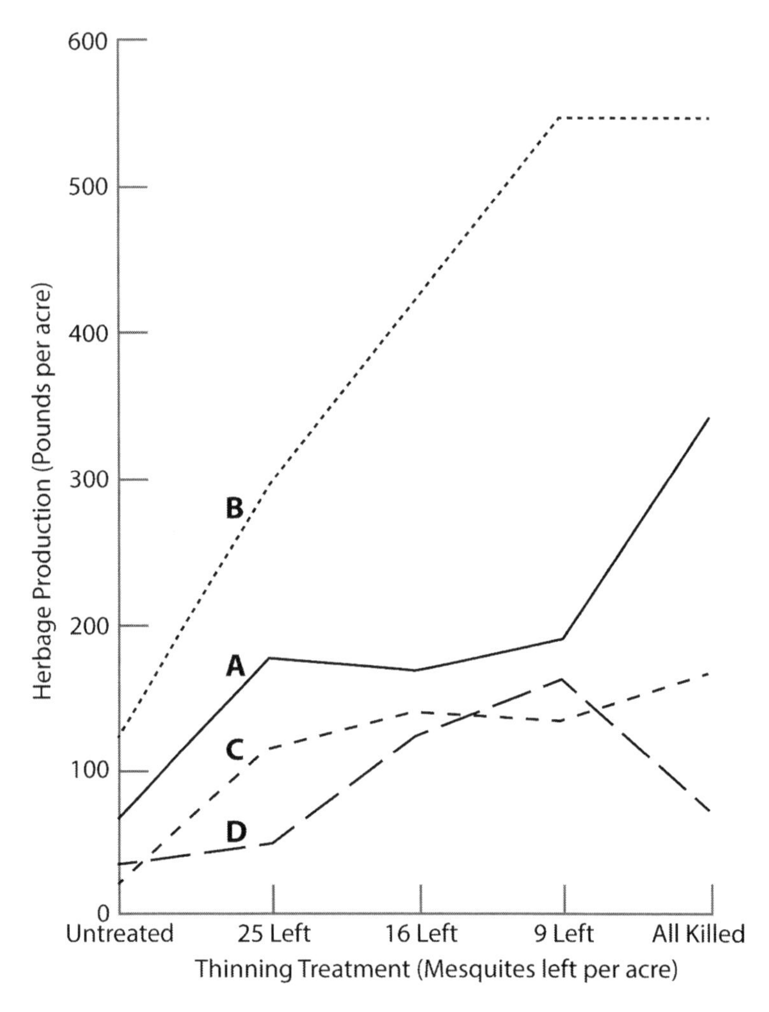

5.7. The effects of mesquite density on forage production following thinning at four sites on the Santa Rita Experimental Range, 1945–1950. Redrawn from Parker and Martin (1952, 46) by Darin Jensen.

an elaborate calculation reminiscent of Jardine's analysis of the costs and returns of fencing some forty years earlier:

> On the 1947 price basis, mesquite control on [an area with 101 mesquites per acre] by use of Diesel oil (the cheapest method) would cost $4.04 per acre or $2,585.60 per section. Mesquite control is estimated to result in an average

> beef production increase of 2,150 pounds per year per section of land. The value of this increased production is $365.07, figured on the basis of $16.98 per hundredweight, the 10-year (1941–50) average selling price of calves at the ranch in southern Arizona. If the investment of $2,585.60 per section for mesquite control is compounded at a 5 percent interest rate, the increased returns realized from mesquite control would liquidate this capital outlay in about 9 years. Assuming that the control work would be effective for 25 years, the benefits of control would be available for 16 years after paying off the investment. (Parker and Martin 1952, 64)

The problems with this calculation are numerous. It was obviously quite sensitive to the costs and prices used as well as the interest rate, all of which could easily change. The site used to calculate the production increase had been treated with arsenic, so the results were presumably greater than would be the case with diesel oil. Moreover, many sites had much higher mesquite

5.8. Applying sodium arsenite to a girdled mesquite tree, Santa Rita Experimental Range, 1942. The method killed mesquites effectively, but it was also expensive and dangerous. Source: National Archives, Forest Service photograph number 425037. By Kenneth W. Parker.

densities—of the four used in the thinning study, for example, only one had fewer than 101 trees per acre, and one had 358—and Parker and Martin had no way of knowing how long the treatment benefits would persist. The economic "analysis" of shrub control, as with fencing before, was wishful at best.

The urgency of shrub control was deemed so great that even fire was tried in a handful of experiments. Where grass was sparse, fuels had to be added: Parker (1939) used stove oil to help burn burroweed, achieving complete mortality; Streets and Stanley (1938) found limited success killing mesquites by burning the bark around the base of the trees with a kerosene torch; and Glendening and Paulsen (1955) placed straw around mesquites to produce more intense fires. Without added fuels, Reynolds and Bohning (1956) burned two plots, 100 acres and 6 acres, in 1952; Cable (1967) burned two 2-acre plots in 1955; and accidental fires in 1950 and 1963 also afforded data (Glendening and Paulsen 1955, Cable 1965). In all these cases, however, mesquite mortality was meager: about half of very small plants (less than half a centimeter in diameter in one study, less than half an inch in another) died completely, but the rest, along with almost all larger plants, resprouted even if their aboveground parts were killed. Benefits to grass production were correspondingly small, short-lived, and/or dependent on unpredictable postfire rainfall patterns. Black grama experienced high mortality, especially compared to Lehmann lovegrass, which thrived with burning—meaning that even if fire did curtail mesquite, it might also negatively impact native grasslands (Cable 1971). In effect, the half century without recurrent fires had allowed mesquites to establish and pass a threshold beyond which the return of fire could not reverse the change; Parker and Martin (1952, 14) therefore judged the relationship of fire to mesquite "a moot question." The overall conclusion researchers reached was that fire posed high risks and low rewards as a shrub control tool, and fire again waned as a research topic.

As with artificial reseeding, shrub control became increasingly a matter of engineering machines and chemicals that could achieve economies of scale adequate to the large extent and relatively low value of rangelands. The CCC had provided the labor for removing plants manually during the Depression, but after the war such methods became prohibitively expensive. Mechanical means of removing unwanted range plants also had a longer history, but in the postwar period, larger, more powerful machines were designed and tested throughout the West. In 1947, the USDA imported a "stump-jump plow" from Australia and set to work modifying it for use on sagebrush rangelands, resulting the following year in the brushland plow, a three-ton implement

5.9. The Bushwacker, attached to a bulldozer. "A heavy-duty machine for clearing land through the disintegration of brush, vines, undergrowth and trees." Photo from promotional flyer found in the National Archives, date unknown.

with seven pairs of disks independently suspended on heavy coil springs to cope with variations in terrain while reducing repairs and lost time (Coldwell and Pechanec 1950). Twice as expensive as an ordinary wheatland-type plow, the brushland plow required a bulldozer to pull it, but it achieved "kills of 100 percent" on rock-free sagebrush rangelands and "90 percent or more on rocky areas" (Sampson and Schultz 1957, 18). It was but one of dozens of implements and techniques developed for brush control by the USDA or private equipment firms. The enormous Bushwacker (figure 5.9), a bulldozer attachment with its own 175-horsepower diesel engine powering a "disintegrator" of twenty spinning steel flails, was capable of destroying trees up to eight inches in diameter at a rate of one acre per hour (Sampson and Schultz 1957, 21). How widely it was used is unknown, but it suggests the lengths to which mechanical inventiveness could go. Simpler, and more common, was "chaining," in which two bulldozers would drag a sea-anchor chain across a swath of rangeland to knock over brush.

Range scientists were only tangentially involved in most of the mechanical brush control research—engineers were more active, and large private ranches such as the King Ranch in Texas also played prominent roles. In its first fifty years of publication (1949 to 1999), the *Journal of Range Management* published only sixty-eight articles on mechanical brush control, with more than three-quarters of them appearing after 1970; even fire research, with eighty-two articles, was published more frequently (Bovey 2001, 24). In part, the low publication rate reflected the fact that, despite all the horsepower these techniques wielded, mechanical control was largely a failure when applied to resprouting shrubs such as mesquite. If the treatment left the root crowns intact, the plants would quickly grow back, often in multistemmed forms that were even harder to eradicate. Bulldozer attachments were developed to solve this problem, too—long steel "stingers" that would penetrate below ground and break off the root crown—but they required the same intensive, plant-by-plant work that had made arsenic application so costly.

Chemicals appeared to offer the only cost-effective means of large-scale eradication, and range scientists were more actively involved in this line of inquiry, publishing 205 articles on the subject between 1949 and 1999, including 114 in the 1970s and 1980s alone (Bovey 2001, 24). First synthesized in 1941, the phenoxy herbicide 2,4-D (2,4-dichlorophenoxyacetic acid) was initially kept secret while the military studied its possible wartime applications. It was applied from the ground on the Santa Rita Experimental Range in 1946, and soon thereafter from the air in combination with a related and more powerful chemical, 2,4,5-T (2,4,5-trichlorophenoxyacetic acid) (figure 5.10). By defoliating broadleaf plants, herbicides such as these could kill shrubs that did not resprout from the roots, such as big sagebrush (figure 5.11), but, as with mechanical control, they were less effective against mesquite: Parker and Martin (1952, 60) characterized the results as "highly erratic. Usually, foliage applications of these chemicals on velvet mesquite have resulted in nearly complete defoliation, partial top-kill, and a very low percent of actual kill." But by the end of the 1950s, a team led by Fred Tschirley had perfected a method of range improvement that combined chemical herbicides, nonnative grass seeds, and aerial application of both (Tschirley and Hull 1959).

In May 1954, ninety acres of the Santa Rita, with an average of 225 mesquite trees per acre, were sprayed with 2,4,5-T mixed with diesel oil and water at a rate of five pounds of herbicide per acre (Cable and Tschirley 1961). On the same day, the area and sixty adjacent acres were aerially seeded with Lehmann lovegrass at a rate of one pound per acre. Livestock were excluded during the summer growing seasons of 1954 and 1955 to assist grass establishment. Although the mesquites were almost completely defoliated by the spray, 95 to 100 percent of them resprouted the following spring, so eighty

5.10. Chemical brush control on the Santa Rita Experimental Range. Top: Treating burroweed with 2,4-D using the "Bes-Kil artificial fog applicator," 1946. Source: National Archives Forest Service photograph 450396. By Kenneth W. Parker. Bottom: Aerial spraying of mesquite with 2,4-D and 2,4,5-T, the combination later known as Agent Orange. Source: Santa Rita Experimental Range archives.

5.11. Rangeland manipulations on the Beaverhead National Forest in Montana, 1963. The caption for this photo reads: "Contractor building range fence for the Rock Island Oil Co. on the Long Creek Allotment. This range was aerially sprayed for sagebrush control in 1963 and is being fenced into pastures for rotation grazing." Source: National Archives, Forest Service photograph 504548. By George R. Wolstad, July 24, 1963.

of the ninety acres were sprayed again in June 1955, some at one-half pound per acre and some at three-quarter pound per acre of 2,4,5-T. Whereas mesquite mortality on the ten acres that had been sprayed only once was just 2 percent, the resprayed areas reached 30 and 54 percent, respectively, in three years, and slightly higher after five years. Native grass production, meanwhile, increased nearly threefold on the sprayed area during the summer following the initial treatment, and was twice as great as on the unsprayed area over the six-year study period. Lehmann lovegrass production went from zero prior to seeding to 424 pounds per acre on the sprayed area, compared to 186 pounds per acre on the unsprayed area. Cable and Tschirley calculated that, at 4 percent interest, the spray-seed-spray treatment returned a profit after the third growing season; after six years, it returned $7.85 per acre more than seeding alone.

The chemical work at the Santa Rita in this period was just a small part of a much larger enterprise. The research that made aerial herbicide application possible had two primary driving forces: the military—which had been developing aerial chemical warfare techniques since World War I (Buckingham 1982)—and crop agriculture—chemical warfare against insect pests having originated in southern California's citrus orchards in the 1880s (Romero 2015). Applying 2,4-D and 2,4,5-T to rangelands was just an extension of their use to control weeds in other settings. In 1961, phenoxy herbicides were sprayed on "about 40 million acres plus hundreds of thousands of miles of roadsides, railroads, and utility rights of way . . . in the United States" (Buckingham 1982, 196). The same year, President John F. Kennedy authorized military use of herbicides in Vietnam, initially to clear vegetation along highways and eventually to kill food crops as well as to defoliate tropical forests. By war's end, some 20 million gallons of herbicides had been sprayed on 3.5 million acres of South Vietnam; four-fifths of the total was applied to inland forests, which, like mesquite, would resprout and therefore require reapplication (Sills 2014, 77–78). Slightly more than half of the herbicides used were an equal-parts mixture of 2,4-D and 2,4,5-T known as Agent Orange, after the color of the barrels in which it was stored. The code name of this massive chemical deployment was Operation Ranch Hand.[6]

The total area of US rangelands affected by artificial improvement is unknown. Mechanical shrub removal, aerial chemical application, and large-scale reseeding were all expensive relative to returns from livestock, and even at their height in the 1960s and 1970s they were only applied to a fraction of US rangelands—albeit a likely total of tens of millions of acres. When concerns about adverse health effects prompted the USDA to restrict 2,4,5-T in 1970 (it was banned altogether in 1985), another round of research was launched with newer chemicals such as paraquat, tebuthiuron, karbutilate, picloram, and glyphosate. A pelleted herbicide application on the Santa Rita Experimental Range in 1979, for example, increased forage production from 250 to 3,200 kilograms per hectare after twenty-eight months (Cox et al. 1982, 18). But over the 1980s, the popularity of the postwar model of artificial range improvement—based on heavy equipment, airplanes, non-native grasses, and industrial chemicals—waned. Tools were available in abundance—a 1980 *Revegetation Equipment Catalog*, published by the Forest Service (Larson 1980), ran to nearly 200 pages—but for numerous reasons, using them no longer made as much sense as it had before. Many ranchers found that the forage gains were not as great or as persistent as they had hoped (Cox and Jordan 1983). Lehmann lovegrass, although preferable to shrubs, was still less desirable than native perennial grasses for livestock production—and without fire mesquites soon began to spread again anyway (McPherson and Weltzin 2000; Archer et al. 2011). Increases in the cost of diesel fuel following the oil crises of the 1970s, coupled with stagnant livestock prices, meant that brush control was no longer economically rational even if it did produce more forage.

In several respects, the prospects for artificial range improvement in the Southwest changed very little in the forty years after Jerome Eddy wrote to his congressmen in 1945. Most of the techniques developed in that time were eventually judged to have failed. In a 1982 review of restoration efforts in the Sonoran and Chihuahuan deserts, a team of prominent range scientists noted several chronic problems stretching back seventy-five years. Studies had lacked onsite weather data, consistency in data collection, replication, and accurate information about seed sources. Many experiments had not been published because they had failed, resulting in needless repetition over time. Selective grazing had resulted in overgrazing of introduced plants even at low stocking rates. Above all, though, the problem was spatial and temporal variability: "Even when seedbed methods and seeded species are the same, results vary from site to site and year to year" (Cox et al. 1982, 15). This was essentially a problem, once again, of scaling up from research plots to the larger range: "The situation becomes more confusing when positive results, obtained at atypical sites or in atypical years, were extrapolated and recommended for use over large areas" (Cox et al. 1982, 16).

Achievements that might have been viewed as successes in Eddy's time came to look like cases of unintended consequences, because the introduced species transgressed the line between artificial and natural, escaping the control that manipulation was supposed to provide. Among ecologists and conservation biologists, several nonnative grasses that were once greeted as miracle cultivars, such as Lehmann lovegrass and Buffel grass (*Centrus ciliarus*) in the Southwest, are now seen as invasive species in their own right. In the Great Basin, the loss of sagebrush habitat—accelerated by a positive feedback between cheatgrass and fire (Knapp 1996)—has emerged as a threat to the region's wildlife; today, whether sagebrush ever actually "invaded" Great Basin grasslands is disputed (Welch 2005). For federal agencies, finally, the growing influence of the environmental movement impinged on artificial reseeding from several directions: chemicals and nonnative grasses became more tightly regulated, and environmentalists' opposition to grazing on federal lands—empowered by legislation such as the National Environmental Policy Act and the Endangered Species Act—made the Forest Service and Bureau of Land Management wary of interventions that appeared to benefit ranchers. Against a backdrop of rapid population growth and urbanization in the interior West, the political and economic power of the livestock sector diminished while demand for recreation, wildlife protection, and other values from public rangelands grew (Sayre 2002).

Even today, however, artificial improvement remains a tantalizing prospect, notwithstanding its past failures. The problem of shrub encroachment and grassland loss has not gone away. Despite hundreds of thousands of acres of shrub control and a steady reduction in stocking rates across the region since about 1920, an estimated 84 percent of perennial grasslands in the Apachean Highlands bioregion of the Southwest (a classification developed by the Nature Conservancy) has been invaded by shrubs; three-fifths of this area is considered beyond restoration, either because grasses are too scarce to carry a fire or because nonnative grasses dominate to such an extent that native grasses cannot compete (Gori and Enquist 2003). Gains in forage and livestock production alone may not justify artificial range improvement (if they ever did), but other goals—wildlife habitat, biodiversity, watershed quality, or recreation, for example—also must be weighed. In the words of a major recent assessment, "When the value of ecosystem goods and services beyond those associated with livestock production are taken into account, a more favorable picture of brush management begins to emerge" (Archer et al. 2011, 149).

US rangelands eluded the sustained and determined efforts of generations of scientists to remake them in the image of their choosing. Ironically, it is for this reason that rangelands persist: no one has found a way to convert them profitably into other kinds of land. The line between natural and artificial improvement thus traces more than the progress and pitfalls of scientific knowledge; it also reveals the changing boundaries of political and economic rationality for investments in manipulating rangelands. Alternative funding sources, such as government programs for restoration, and potential revenue streams—biofuels from juniper trees, for example, or payments for carbon sequestration or other ecosystem services—may yet make manipulation economically rational (Havstad et al. 2007). Meanwhile, what counts as success scientifically has also changed. Range scientists have long known that reducing livestock grazing would not reverse shrub encroachment once it had begun, but only recently have they begun to renounce the ideal of restoring "original" conditions in favor of a notion of "remediation" that relies on measures of ecosystem processes rather than forage production or species composition (Herrick, Havstad, and Coffin 1996). This reflects, in part, a loosening of the lock grip of Clementsian ecological theory since the 1980s, as we will see in chapter 7. The non-equilibrium alternative displaces climax conditions as the benchmark and goal of rangeland management, shifting expectations of what is possible and therefore what the goals of improvement should be. "Because grazing management alone is generally not sufficient to curtail or reverse shrub encroachment, progressive brush management is a potentially important tool for grassland conservation" (Archer et al. 2011, 109). If a singular climax does not exist, and succession is not a universal natural tendency of all plant communities, then in some cases intensive manipulation might yet again be expected to "pay real dividends."

SIX

The Western Range Goes Global

Neo-Malthusianism and Pastoral Development

From about 1910 to 1948, as range science was developing into a discipline, forces from outside the United States had little apparent effect on the field. But before and especially after that period, international influences were significant. It was in the British colonies of Australia and New Zealand, beginning in the 1840s, that expansive rangelands were first divided into allotments for exclusive lease to private livestock operators, with fees based on the estimated number of livestock an allotment could support (Shear 1901, 59–61). The systems there appear to have prompted the application of the term *carrying capacity*—which was originally coined in connection with steam-powered shipping—to rangelands (Sayre 2008), and they helped inspire the first public land grazing leases in the United States, established by the state of Texas in 1883, which in turn informed the lease systems on national forests and Bureau of Land Management lands. Australian experiences with rabbits and dingoes also provided precedents for rodent and predator control (Palmer 1897b), and the use of fencing to exclude predators there and in South Africa may have helped to inspire the Coyote-Proof Pasture Experiment (Lantz 1905). The British influence on US rangelands is hardly surprising given the prominence of British investors, stockmen, and livestock breeds in the early western ranching industry (Sayre 1999)—Francis S. King, who advised Frederick Coville about fencing in 1907, is a case in point. And, as we have seen, plants from around the world played important intended and unintended roles in US rangelands and range science throughout the twentieth century.

Interactions between US range science and rangelands overseas intensified in the postwar period, first from the United States outward and subsequently in the other direction. Despite the importance of range livestock production in nations around the world, only in the United States had it given rise to an institutionalized science.[1] International development agencies perceived rangelands as "unimproved" and therefore ripe for "modernization," and US range science provided the concepts, evidence, and expert authority from which "pastoral development" unfolded. Beginning in the late 1950s, the World Bank, the US Agency for International Development (USAID), and other bilateral and multilateral agencies invested billions of dollars to increase range livestock production in the so-called Third World, and the model they sought to implement was the one that had emerged over the preceding half century in the United States. It conformed well to the developmentalist agenda: dedicated to commercial production and increasing efficiency and output; reliant on investments in "improved" breeds and technologies such as fencing, wells, and veterinary medicine; and imbued with a conviction that expert knowledge could liberate local people from the premodern ideas that putatively held them in a state of backwardness and poverty.

The political currents into which range science stepped in this new context were much more powerful and lopsided than range scientists had ever experienced at home. From Anglo-American settlement through most of the twentieth century, ranchers enjoyed disproportionate political power in the western United States, and institutionally, range science both reflected and relied on this fact. If ranchers and environmentalists only ever reached a kind of stalemate in the late twentieth century, it was because their political power and legal rights were roughly equal, with neither side able to dislodge the other from public lands as a whole. Pastoral development, by contrast, was an international enterprise with enormous disparities of power and wealth between the United States and development agencies on the one hand and the nations of the developing world on the other, and these disparities were further compounded by the fact that pastoralists were often among the most disempowered members of those nations, marginal to both colonial and postcolonial elites.

The ideas of US range science were quickly swept up into these asymmetries. The concept of carrying capacity, for example, resonated perfectly with the neo-Malthusian fears of overpopulation and natural resource degradation that gave international development such breathless urgency during the Cold War. When they first lent their expertise to development agencies, range scientists could not have foreseen that their central claims, based on a Clementsian interpretation of US rangelands, would provide the raw materials from which much broader and more politically charged assertions about "desertification" and "the tragedy of the commons" would be discursively constructed and mobilized. By the late 1960s, the world's human population was routinely discussed using parables and analogies of livestock in a pasture, and the imperative to control human population growth—coercively if necessary—was cast as akin to imposing correct stocking rates to prevent catastrophic overgrazing.

Carrying Capacity, Overpopulation, and the Cold War

In 1948, one year after Bernard DeVoto had first excoriated ranchers in the pages of *Harper's* and the same year that the Society for Range Management was founded, two best-selling books energized the American conservation movement: William Vogt's *Road to Survival* and Fairfield Osborn's *Our Plundered Planet*. "Both books articulated a shift in conservation thinking toward interconnection and ecology, overpopulation, overconsumption, limits and sustainability—and potential apocalypse" (Robertson 2012, 47). Vogt had studied ecology and ornithology, worked at a bird sanctuary after college, and performed reconnaissance work in Latin America for the Office of the Coordinator in Inter-American Affairs during the war. He would later serve as national director of the Planned Parenthood Foundation of America from 1951 to 1961 and as secretary of the Conservation Foundation (precursor of the World Wildlife Fund) from 1964 to 1968. Osborn was the grandson of railroad tycoon William Henry Osborn and the son of Henry Fairfield Osborn, a famous geologist, paleontologist, and eugenicist; he had studied biology at Cambridge before going into business, and in 1935 he retired, at the age of forty-eight, to devote himself full-time to conservation causes as the well-connected president of the New York Zoological Society and, later, the Conservation Foundation.

Vogt and Osborn were instrumental in the rise of neo-Malthusianism into mainstream American popular and political thinking in the run-up to the Cold War, and the vehicle they employed—Osborn implicitly, Vogt expressly—was the idea of carrying capacity (Robertson 2012). Both were friends of Aldo Leopold, who had spent an unhappy but seminal nine months working for the Office of Grazing in the Forest Service's southwestern regional office in Albuquerque in 1914–1915. It was there that Leopold had first encountered the idea of carrying capacity, a "discovery [that] would reverberate through his work for the rest of his life" (Meine 1988, 136). He developed the concept in relation to wildlife in his landmark 1933 textbook, *Game Management*, and in a lecture to his students at the University of Wisconsin in March 1941, with World War II under way, he speculated that war might be a mechanism by which human populations were ecologically self-limiting, just as predation or habitat conditions limited populations of game species in the wild (Leopold 1992). Whether Leopold was inspired to make this leap by Vogt and Osborn, or the other way around, is unknown, but both Vogt and Osborn "drew a straight line from resource scarcity and degradation to war" (Robertson 2012, 49), and both used Leopold's work to illustrate the central argument of their books: like an irrupting population of deer or rabbits, humanity was degrading the world's natural resources and outstripping Earth's carrying capacity.

The work of population biologists Royal Chapman and Raymond Pearl, who had refined the idea of carrying capacity in controlled laboratory experiments, also informed Vogt's and Osborn's ideas. Based on his work with flour beetles, Chapman (1928, 114) had written that, "it seems evident that we have in nature a system in which the potential rate of reproduction of the animal is pitted against the resistance of the environment, and that the quantity of organisms which may be found is a result of the balance between the biotic potential, or the potential rate of reproduction, and the environmental resistance." This formulation became Vogt's way of defining carrying capacity, as the "bio-equation" C = B: E, in which C stood for carrying capacity, B for biotic potential, and E for environmental resistance. Biotic potential, Vogt wrote, had "an absolute or *theoretical* ceiling that is never reached, except under extraordinary conditions," and "a very large number of *practical* ceilings," which were "in most of the world dropping lower every year. . . . The practical ceiling is imposed by the *environmental resistance,* which is the sum of varying but always great numbers of limiting factors acting upon the biotic potential" (Vogt 1948, 16, emphases in original). Chapman and Pearl's experiments lent empirical support and scientific credibility to the idea that *species* had inherent carrying capacities—densities they approached but could not exceed even under ideal environmental conditions—but this only appeared to resolve the tensions that afflicted carrying capacity in range science. Like James Jardine's forage acre, biotic potential was an abstract idealization, and Vogt's equation was tautological: environmental resistance was a concept necessitated by postulating a static, ideal carrying capacity from which reality always and necessarily deviated. As in range science, carrying capacity for neo-Malthusians was simultaneously ideal and real, prescriptive and descriptive, normative and positive—but here it was applied to humans instead of livestock and oriented toward an apocalyptic future rather than an original, pristine past.[2]

Concurrently with the publication of their books, Vogt and Osborn convened or addressed major international conferences that drew together academics and government officials from across the United States and overseas to discuss natural resource issues. As head of the conservation section of the Pan American Union, Vogt served as secretary general of the Inter-American Conference on Conservation of Renewable Natural Resources (IACCRNR) in Denver, Colorado, in September 1948, sponsored by the State Department. The conference proceedings opened with "A Preliminary Statement," presumably authored by Vogt, which likened humanity to livestock in a global pasture. "World populations have now increased until there are only about 2 acres—less than one hectare—of productive land for each individual and, while the destructive practices mentioned above are daily causing these 2 acres to shrink, populations are mounting at the rate of about 50,000 people per day" (IACCRNR 1948, 71). Osborn delivered a speech entitled "Action Must Be Taken," in which he declared: "We would not be gathered here did we not realize that the New World is new no longer. From Canada to Patagonia there are too many evidences of land deterioration. The tempo of land misuse is quickened by rapidly increasing populations throughout the Western Hemisphere, as in truth throughout the world." World population was increasing by half a million people every two months, he noted. "A continuation of the present rates of depletion can lead only to disaster" (IACCRNR 1948, 507–8).

Other conference speakers included such academic luminaries as demographer Kingsley Davis, anthropologist Julian Steward, and geographer Gilbert White; government officials such as Hugh Hammond Bennett and Marion Clawson, heads of the Soil Conservation Service and the Bureau of Land Management, respectively; and range professionals such as W. R. Chapline, F. G. Renner, and Arthur Sampson. Chapline's address stressed the improvements that could be made to rangelands through the application of scientific research. "Improved practices, developed by research, have already brought millions of dollars of savings and increased revenue to stockmen annually. Further developments promise even greater contributions in years to come" (IACCRNR 1948, 391). Borrowing numbers from *The Western Range,* Chapline asserted that historic overgrazing had reduced the carrying capacity of western US rangelands by nearly half. Fortunately, scientific investigations "furnish knowledge of the underlying factors and principles involved in range restoration and use," including "a thorough understanding of the climatic, soil, moisture, and plant relationships, the physiological reaction of different plants to grazing, and the influence of all these on the successional and other ecological trends of the vegetation" (IACCRNR 1948, 392). Chapline listed nine areas in which science had informed management for the betterment of US rangelands, including grazing capacities, trend and condition measurements, rotational grazing, noxious plant eradication, and range reseeding. He closed with the claim that "Land with very little actual grazing value but which might have sold for $1 an acre has, through reseeding that cost $3 to $5 an acre, increased $10 to $12 or more in value" (IACCRNR 1948, 396).

The most impassioned statements regarding rangeland conservation were delivered not by a range scientist but by the famous botanist (and retired president of the University of Arizona) Homer Shantz, who had earned his PhD in 1905 at the University of Nebraska under the tutelage of Charles Bessey and Frederic Clements. "The grasslands of the world, especially those skirting the semiarid regions, have been so overused that in many regions practically none of the natural vegetation remains. All that survives of the original flora is the less palatable weeds and shrubs which cling to the land in an endeavor to maintain a soil cover" (IACCRNR 1948, 149). Shantz had worked for the Bureau of Plant Industry from 1908 to 1926 and served as lead botanist for the Smithsonian Institution's African Expedition of 1919–1920, collecting plant specimens from the Cape of Good Hope to Cairo. At the conference, he emphasized the necessity of scientific expertise and a successional lens to perceive and understand grazing impacts: "A trained observer can detect the changes in the cover of grazing lands caused by even a light stocking of domestic livestock and herbivorous game animals. The most palatable plants are eaten first and are the first to be extirpated. The least desirable, the unpalatable plants, take their place. Gradual deterioration may not be detected by an ordinary observer. The total amount of foliage may even be increased but the telltale damaging effect is evident to an experienced eye. Ecological perception will enable the manager to head off many disastrous trends" (IACCRNR 1948, 150). In the absence of such expertise, "Uncontrolled use of forage soon destroys the resource" (IACCRNR 1948, 151). Likening human populations to livestock and wildlife, Shantz intoned ominously:

> Those not versed in range management often assume that as long as the condition of the animals is satisfactory the resource of forage is not overused. . . . A deer or elk population does not level off automatically at the proper level as the resource of forage is exceeded by the demands of an over-numerous herd. It drops by starvation to less than a fourth the number that could have been maintained had the population been properly controlled. It would seem reasonable to expect a similar response in human populations, and that the resources would be largely exhausted before the effect would show in decreased birth rate or increased death rate. (IACCRNR 1948, 152)

The following year, Osborn was one of two keynote speakers at the three-week United Nations Scientific Conference on the Conservation and Utilization of Resources (UNSCCUR), held at Lake Success, New York. President Harry Truman had proposed the conference, declaring that science allied with conservation would "become a major basis of peace . . . [and] world prosperity" (UNSCCUR 1950, vii). In his keynote, Osborn struck a very different tone: "Throughout most of history mankind's record in the handling of renewable resources has been one of failure" (Osborn 1950, 14). "It is certainly no exaggeration to say that the future of mankind will depend upon the degree to which natural resources can continue to be available. . . . The solution has been found in the past principally through exploitation of new regions, even new continents. These exist no longer. . . . Unquestionably the greatest factor of change is the explosive upsurge in population in virtually all countries. . . . The habitable and cultivable regions of the earth are now largely occupied, leaving certain tropical regions and arctic regions as the last remaining frontiers" (Osborn 1950, 13).

Arthur Sampson and Clarence Forsling were among the US experts on a conference panel on "Condition of Grazing Lands." Sampson's paper outlined the successional framework for range management, and he offered two reasons for range degradation in the United States: "the economic and political conditions which induced almost universal overstocking of the range, and secondly, failure on the part of operators and administrators alike to recognize or understand the indicators depicting the state of health of the ranges" (Sampson 1951, 511). Forsling went much further. He was a range scientist and career administrator who by 1949 was special assistant to the secretary of interior. He had graduated from the University of Nebraska in 1915 and then worked for the Forest Service for nearly three decades: as head of the Jornada Experimental Range, the Great Basin Experiment Station, and the Intermountain Forest and Range Experiment Station, and finally as assistant chief of the Forest Service in charge of research from 1937 until 1944, when he left to lead the Grazing Service in the Department of the Interior. In short, he knew both the science and politics of US rangeland management and administration as well as anyone could, and he had experience in both the Southwest and the Great Basin.

Forsling's paper was a masterful piece of selective bureaucratic self-representation, managing simultaneously to take credit for the successes of US range science, turn its failures into opportunities, and claim the authority to prescribe best practices to the rest of the world. After briefly underscoring the significance of grazing lands to world livestock production, Forsling began by declaring that these lands were not producing up to their potential:

"Because of poor husbandry, almost everywhere in the world they are failing to contribute their full share in the production of food and other animal products." Until very recently, science had neglected rangelands, viewing them as too marginal to warrant attention, but Forsling averred that this had finally changed. "No resource has longer been neglected in world agriculture or stands more in need of conservation management for sustained production. It is only in the last few decades that man has begun to realize that grazing land use is far more than a primitive form of agriculture. He has only now begun to explore the problems, and gain the necessary knowledge to develop and apply grazing management with the view of achieving sustained production" (Forsling 1951, 500). Rangelands in the western United States had been severely degraded prior to the advent of range science—like Chapline the year before, Forsling invoked figures from *The Western Range*—and they were still actively degrading in India, China, southern Africa, Australia, and New Zealand. Rangelands around the Mediterranean had been severely and permanently depleted centuries ago, providing sobering illustrations of what was at stake. Everywhere, poor livestock husbandry led to erosion, which reduced productivity, leading to more overgrazing and more erosion in a spiral of degradation. Fearful visions of soil erosion and the collapse of ancient civilizations had animated an American apocalyptic narrative form for two decades (Bennett and Chapline 1928, Sears 1935); by hitching the form to livestock grazing, Forsling cast range science as the savior.

The details varied from place to place, but Forsling claimed that all cases of rangeland degradation could be traced to one or a combination of five causes: overstocking, drought, common land tenure, fire, and invasive plant species. Conservative stocking rates were necessary, such that the most palatable plants were not grazed too much, "even though it may result in but light use of the species of lesser palatability and the leaving of what appears to be a considerable quantity of the growth of the key plant species"; echoing Shantz, Forsling stressed: "The condition of the grazing animals is a false and dangerous guide" (Forsling 1951, 501). Especially in arid and semiarid regions, drought would drastically reduce forage production, and "if grazing is not immediately reduced in proportion to the decline in forage production, there will be extremely heavy overgrazing and injury to the forage stand." The absence of exclusive land tenure was among the greatest challenges of proper management in many countries. "Uncontrolled and unattended grazing lands upon which landless owners of livestock may pasture their animals at will, generally have suffered heavy damage or virtual destruction. Such irresponsible use of land almost inevitably leads to over-use and deterioration. Too many people seek to run more livestock upon the land than it can support and come and go with their herds as they choose. Because of the continued and indiscriminate grazing, the over-used areas never have the opportunity to recover." Forsling acknowledged the political difficulties of imposing exclusive tenure, but he insisted that it was necessary if "continued impoverishment of the land" was to be prevented (502). Regarding fire, Forsling conceded that "Properly controlled fire undoubtedly has a place in coping with certain problems," but science had not yet figured out the details, and until then, "Indiscriminate burning should be barred as one of the worst enemies of grazing lands" (502–3). Finally, excessive grazing could trigger the "invasion of unpalatable species," such as sagebrush in the Great Basin and mesquite in the Southwest. Hence the need for "artificial means such as grubbing, chemical treatment, or controlled burning . . . from time to time to destroy or control the spread of unwanted species" (503).

Forsling drew directly on southwestern range research findings, neatly avoiding any reference to the failures that were both evident and ongoing at the Jornada and the Santa Rita Experimental Ranges:

> Studies have shown that the semi-arid and arid grazing lands which are subject to wide fluctuations in rainfall in the Western United States, should not be stocked in excess of 65 to 75 per cent of the average grazing capacity over a period of years. Such stocking will ensure an ample supply of forage, circumvent serious injury to the range, and avoid serious death losses from starvation or sacrifice of the breeding herds in all but the especially serious droughts. The leeway in the average or better years will provide opportunity for restoration of the forage cover that has been diminished by drought. (Forsling 1951, 502)

Forsling's interpretation was precisely the one that had been institutionalized both within the Forest Service and in range science more generally. The cause of range degradation was overgrazing; imposing limits on stocking was the necessary precondition for sustainable production; only scientists could detect degradation and discern the proper limits early enough to prevent harm; management informed by science could ensure sustainable production and even increase it. The problems that range scientists had encountered in the Southwest did not invalidate these claims but reinforced them by showing how dire the consequences of overgrazing could be. By extension, they also showed how great was the need elsewhere for the knowledge provided by US range science, because most of the world's rangelands were arid or semiarid, like the Santa Rita and the Jornada. "The present condition of grazing lands offers a vast opportunity, as well as need, for measures to stabilize or to increase livestock production in many countries," Forsling concluded. The obstacles to proper management could be great: "In some situations, as for example where nomadic grazing has become established, it may involve the readjustment of the entire land tenure and land use systems," and with the imposition of conservative stocking rates, the small producer "may find that for a period he will be reduced to less than a subsistence level." But Forsling's confidence was unqualified. "During the past five decades, and especially in the last twenty to thirty years, it has come to be realized that large areas would find their greatest economic use for grazing and that they can and should be improved and managed for that purpose" (Forsling 1951, 504–5).

The combined effect of the addresses by Chapline, Sampson, Shantz, and Forsling, among others, was to universalize the experiences of US range science and administration. Findings produced at the scale of plots and pastures at discrete times and places in the United States were extrapolated to all rangelands at all times. To do this required both a highly selective representation of the knowledge in question and an erasure of the historical and geographical specificity of that knowledge. The open range period, when land tenure had been a free-for-all and a flood of capital had washed across the nation's grasslands leaving severe overgrazing in its wake, now stood in for all rangelands that lacked exclusive tenure arrangements; outside capital was seen not as a cause of degradation but as a preventive or remedial solution to it. The successional model of rangeland ecology applied everywhere, although it hailed from the Great Plains and higher elevation rangelands that enjoyed greater and/or more reliable precipitation than many rangelands elsewhere. The more arid and variable portions of the western United States, where the successional theory had not worked well, and which were better analogues for drier rangelands in the developing world, were viewed not as evidence that the theory might be flawed but as proof of the urgent necessity of applying the model to avoid catastrophic degradation. Finally, all these ideas were presented as the product of specifically *scientific* inquiry, which both guaranteed their general validity and disqualified other kinds of knowledge from consideration. That American ranchers lacked experience in US rangelands had been one reason scientific investigations were seen as necessary circa 1900: there had been no pastoralism in North America prior to the arrival of Europeans and their livestock, the indigenous people had been decimated and removed in any event, and Hispanic livestock producers had been ignored (or worse) despite their experience. Now, even pastoralists who had been raising their animals for thousands of years in places such as Africa were judged ignorant of the true, expert knowledge that had been produced in the United States in the preceding half century.

Pastoral Development, Desertification, and the Tragedy of the Commons

If they had been looking for them, postwar development agencies might have drawn cautionary lessons from earlier attempts to turn pastoralists and their rangelands into ranchers and ranches. British investors had started African Ranches, Limited, in Nigeria in the 1910s (Dunbar 1970), and the French had tried to raise European breeds of sheep and cattle commercially in the Sahel in the 1920s and 1930s (Riddell 1982), but all these enterprises had failed. There were also examples in the United States: the Bureau of Indian Affairs and the Soil Conservation Service had tried to apply the ranching model on several Native American reservations—most infamously the Navajo, where huge numbers of sheep and goats were slaughtered to match stocking rates to carrying capacities in the 1930s—and these efforts, too, had been tragic failures (White 1983; Weisiger 2009).

In the 1950s, however, science was associated with progress and success, not failure, and the imperative to increase food production in the developing world took on national security implications. Rangelands were viewed as a major opportunity precisely because they were marginal and "unimproved." The editors of the *Proceedings of the Sixth International Grassland Congress,* for example, of which Chapline and Renner were two, opened their 1952 volume with the following proclamations:

> The battle against hunger and privation is the most important task facing the free nations of the world. The specter of hunger stalks the world. Hundreds of millions of people must exist on diets deficient in proteins and total calories. Hunger breeds discontent and willingness to accept strange doctrines. Hunger weakens the will to resist aggression. Increasing population is exerting steadily mounting pressure on food production capacities. . . . We must find new resources for production. We must reverse the downward trend in soil productivity. And for the attaining of both of these necessary objectives, grassland farming is the most effective weapon in agriculture's arsenal. Grasslands of the world are the largest undeveloped resource for increased agricultural production. More than half of the total land surface of the earth is in grazing lands. Most of these enormous acreages are unimproved. Improvement

> practices, based on research findings, can result in vast increases in production and in the utilization of livestock feed from these grasslands. (Wagner et al. 1952, v)

As pastoral development ramped up through the 1960s and 1970s, knowledge about rangelands had to be translated from its places and institutions of origin, predominantly in the United States, to other places and contexts. Prominent US range scientists such as Harold Heady (Sampson's successor at Berkeley) helped to internationalize the discipline, whether as consultants to development agencies, keynote speakers at conferences, or advisors for foreign students sent to the United States for professional training. The agencies, in turn, hired staff with the requisite training to devise and implement plans. Heady wrote a short textbook entitled *Range Management in East Africa* for the Kenya Department of Agriculture, in which he conceded that fencing, although desirable, was not practical "until the African can be taught the more basic livestock and land management practices and until individual land holdings, game conditions and cash income warrant" (Heady 1960, 92). He also allowed that fire, as well as mechanical and chemical methods, might be necessary for brush control and that "Carrying capacities are extremely difficult, if not impossible, to determine accurately" (39). Otherwise, though, his lessons hewed closely to the standard model of range science, and he concluded by remarking: "All aspects of the social structure must be improved. Probably the first step in the process is to break the tribal customs through decree and education . . . a firm hand is needed to initially force the break and later to guide the action" (119–20).

A key document in the mobilization of range science for pastoral development personnel was the United Nations Food and Agriculture Organization's (FAO's) 1967 *East African Livestock Survey*. It is worth examining in detail for evidence of how development experts understood rangelands and pastoralists, what they sought to do, and the lengths to which they felt justified in going to realize their ambitions. The director and main author of the *Survey* was Sir Donald MacGillivray, whose passing shortly before its publication was noted in a box on the back of the title page. "The reader will appreciate the challenging inspiration in his work," it read. Although the *Survey* officially covered only the nations of Kenya, Tanzania, and Uganda, the box went on to say: "This report will serve as a pattern and a textbook of animal industry to all developing countries in tropical Africa."

In the *Survey*, the transition from pastoralism to ranching was framed as a multifaceted but teleological process: "The progression of an industry from (say) the nomadic herding of unimproved cattle on indigenous vegetation to the sophisticated techniques of ranching genetically superior stock on improved pastures can be regarded as a continuous evolution, in which the passage from one stage of development to the next is almost imperceptible. . . . These improvements may be technical (e.g. disease control), social (e.g. new systems of land tenure), [or] economic (e.g. credit facilities or improved marketing organization), but they must occur in an appropriate order in relation to one another if progress is to be continuous and as rapid as possible" (FAO 1967, 19–20). Above all, the *Survey* was optimistic: "Although it is true that many grazing, destocking and water control schemes have failed in the past, there are many encouraging signs that some pastoral tribes at last are wanting to develop new methods of husbandry and new ways of life" (26). Environmental conditions on East African rangelands were marginal, the *Survey* conceded, "but it must not be assumed that the environment of much of East Africa is so bad, and the standards of livestock management so poor, that the only thing to do is to find a breed hardy enough to produce milk or meat in spite of them. Future research will show how the environment can be improved, and training and extension activities are already resulting in substantial improvements in livestock management" (137). All told, there were about 15 million acres in East Africa "which now await development; this area is capable of running on average about one beast to 15 acres," meaning that about 1 million additional cattle could be envisioned in the region (31). This was the gap between actual and potential carrying capacities that modernization aspired to close.

There were myriad environmental challenges involved in East African pastoral development, the most notorious being tsetse flies, the vector of the deadly livestock disease trypanosomiasis and human sleeping sickness, which traditional pastoralists had long since learned to avoid by keeping their stock away from infested areas. But the *Survey* saw the pastoralists themselves as the most frustrating obstacle to progress. Currently unoccupied areas that might be rendered usable for livestock—by eliminating tsetse, for example, or by drilling wells in waterless areas—were particularly attractive precisely because "no time and effort needs to be spent in modifying established customs and attitudes, particularly those relating to land tenure and to livestock. However, it is still necessary to inculcate good management practices into new settlers by rigid control until they become habitual" (FAO 1967, 27). The *Survey* pointed to successful modern ranches owned and operated by European settlers as proof that environmental obstacles could be overcome (41).

If the core tenet of proper animal husbandry was strict control of livestock numbers, a major problem was pastoralists' tendency to treat livestock as the endpoint or goal of production rather than as a means to the end of obtaining cash incomes. The "cattle complex" interpretation of African pastoralism, made famous by anthropologist Melville Herskovits (1926), was frequently cited in development reports as proof of the natives' economic irrationality (FAO 1967, 29). Without greater integration into the commercial economy, pastoralists would simply hold on to their animals, accumulating as many as they could. "Efforts to raise the imaginative horizons of the pastoral tribes must concentrate upon loosening their traditional regard for livestock. . . . Only when he has to pay cash for school fees, taxes, or his limited range of consumer goods . . . does the pastoralist need to sell an animal and obtain a different currency. If his appetite for goods valued in money could be cultivated, his willingness to sell cattle, sheep or goats, would be correspondingly increased" (20). Increasing dependency on market exchange had the added benefit of inducing a greater willingness to participate in formal education: "Experience has shown that resistance to education becomes progressively less as people begin to appreciate the value of consumer goods and begin to patronize, if only spasmodically, the local cattle market and trading centers" (26).

Most problematic of all was the pastoralists' adherence to communal land tenure. "Experience throughout East Africa has repeatedly demonstrated that attempts to improve livestock productivity have only been rewarded when traditional systems of communal land tenure have been replaced by recognized holdings, the legal titles or sole rights of use being held by individuals, small family groups, co-operatives, companies or corporations" (FAO 1967, 21–22). Without exclusive tenure to land, none of the rest of the tools of range science—fencing, controlled stocking rates, rotational grazing schemes, water development, disease control, improved breeding—could be implemented with any chance of success, not only for practical or logistical reasons but because the capital to pay for modern methods required it. Tenure reform was "the prerequisite for the investment of private savings and personal effort" and necessary "before long-term loans can be obtained from banks and other credit organizations" (22). "Of all the factors that may affect the development of the livestock industry favorably or unfavorably none is more important than the form of land tenure in force in a particular area. . . . One can in fact observe that without the right form of land tenure livestock improvement becomes virtually impossible" (38).

Range science needed land in the sociospatial form of the range, and the *Survey* was not shy about the use of coercive state force to compel pastoralists to conform to it. "Certain measures must be enforced in the pastoral areas as much for the good of the nation as a whole as for the good of the local people: for example, there must be strict veterinary supervision of stock movements, marketing and quarantine; large-scale disease control and (where feasible) eradication campaigns must be mounted and compulsorily maintained on a virtually permanent basis. Moreover the local people cannot be allowed to destroy the natural resources of soil and vegetation; and therefore, when such destruction appears probable, soil conservation measures must be compulsorily introduced" (FAO 1967, 21). One mechanism of compulsion would be debt. "The planned development of new country necessarily involves a high capital expenditure," and loans would therefore be needed. "The terms governing such loans should be worded to ensure that all participants practice a system of enlightened husbandry . . . and, if necessary, to ensure that the instructions of Government staff are promptly and strictly carried out" (27). The high variability of rainfall, which otherwise posed severe environmental challenges, was seen as advantageous for social reform: "The rate of development may be considerably accelerated in times of economic and nutritional stress, such as may be occasioned by a prolonged and serious drought . . . it is likely that a community which has experienced starvation and serious losses among its livestock will no longer shrug off its hardships as 'shauri la mungu' [Swahili for "act of God"]. Experience has shown (e.g., in Kajiado Masailand) that such people are prepared radically to alter their traditional customs, land tenure systems and husbandry practices once the threshold of co-operation is reached" (26). The *Survey*'s ultimate rationale for such coercive measures—at best opportunistic, at worst inhumane—was the need to increase production to match the growth of the human population. "While a considerable effort should be made to increase agricultural, livestock and industrial production, an even greater effort is needed to restrict the present excessive rate of growth of the population." The development measures being proposed "cannot be realized while the population is allowed to expand at its present rate of about 3 percent per annum" (29).

A much more widely read justification for coercive population control measures appeared the following year in the pages of *Science*: Garrett Hardin's (in)famous article entitled "The Tragedy of the Commons." The core narrative device of Hardin's argument was a parable:

> Picture a pasture open to all. It is to be expected that each herdsman will try to keep as many cattle as possible on the commons. Such an arrangement may work reasonably satisfactorily for centuries because tribal wars, poaching, and disease keep the numbers of both man and beast well below the

> carrying capacity of the land. Finally, however, comes the day of reckoning, that is, the day when the long-desired goal of social stability becomes a reality. At this point, the inherent logic of the commons remorselessly generates tragedy. As a rational being, each herdsman seeks to maximize his gain. (Hardin 1968, 1244)

Because each additional animal benefits the herdsman more than the animal's impacts on the common pasture harms him, "the rational herdsman concludes that the only sensible course for him to pursue is to add another animal to his herd. And another; and another. . . . But this is the conclusion reached by each and every rational herdsman sharing a commons. Therein is the tragedy. Each man is locked into a system that compels him to increase his herd without limit—in a world that is limited. . . . Freedom in a commons brings ruin to all" (1244). Ironically, Hardin portrayed US rangelands—which were *not* commons and hadn't been for decades—as further proof of his argument. The tragic logic of the commons, he wrote, "is understood mostly only in special cases which are not sufficiently generalized. Even at this late date, cattlemen leasing national land on the western ranges demonstrate no more than an ambivalent understanding, in constantly pressuring federal authorities to increase the head count to the point where overgrazing produces erosion and weed-dominance" (1245).

Hardin's thesis about land tenure and overgrazing was hardly new—much the same argument can be found in the earliest reports on rangeland degradation in the western United States (e.g., Bentley 1898; Smith 1899; Richards et al. 1905; Wooton 1908) and in the work of Clements (1920). But those writers had advocated leasing and fencing the West's rangelands to enable control of stocking rates, whereas Hardin used the overgrazed pasture as a metaphor for humanity's burden on the planet. The spatial scales were wildly different, as were the practical implications—for Hardin the point was "the necessity of abandoning the commons in breeding" (Hardin 1968, 1248). But the premise in both cases was the same: only exclusive possession of resources can prevent overexploitation. The institution of private property may be unjust, Hardin conceded, but "The alternative of the commons is too horrifying to contemplate. Injustice is preferable to total ruin" (Hardin 1968, 1247).

Appearing in the world's premier scientific journal at a time when neo-Malthusianism was at its peak of popularity in both the foreign policy establishment and the nascent environmental movement, Hardin's thesis gained wide currency. The seeming relevance of its central parable to pastoral development was inescapable, strongly reinforcing views that were already prevalent among development experts regarding rangelands and pastoralists. The subsequent two decades were "a period of large-scale development projects by the international donor community, many of whose policies, driven by the 'tragedy of the commons' thesis, emphasized privatization of the range, commercial ranching, and sedentarization of nomads, particularly in Africa" (Fratkin 1997, 236). The customary systems that pastoralists around the world had developed for allocating access to land without recourse to private property were overlooked or dismissed as anachronistic.

Both the *East African Livestock Survey* and Hardin's "Tragedy of the Commons" appeared just as a severe drought was beginning across the African Sahel. Between 1968 and 1973, about 4 million head of livestock and 100,000 people perished from disease and starvation, despite massive shipments of food relief by USAID and other agencies (Cohn 1975). Images of emaciated Africans were seared into the minds of Westerners in what became one of the world's first televised humanitarian crises. Although scholars soon blamed the famine less on drought than on political and economic dislocations traceable to the colonial period—export crop production had in fact increased during the drought (Lofchie 1975)—the prevailing view as the crisis unfolded held that overpopulation and overgrazing were the primary causes: modern medicine and technology, combined with relatively good rainfall in the decades preceding the drought, had increased both human and livestock numbers beyond the carrying capacity of the region (Grove 1974, Cohn 1975). The drilling of deep borehole wells to support livestock in dry times and places—a widespread component of pastoral development projects in the region—was now seen as a contributing factor to the increase in livestock numbers (Cohn 1975).

It was in this context that a term coined by the French in colonial North Africa—*desertification*—burst on the anglophone world as the latest way to conceive of the crisis. A 1972 USAID report estimated that "about 250,000 square miles of arable land (i.e., suitable for agriculture or intensive grazing) has been forfeited to the desert in the past 50 years," with the sands of the Sahara advancing southward at rates as high as thirty miles per year. "*Overgrazing* by livestock appears to be a major factor in desert encroachment" (USAID 1972, 2, 4, 9, emphasis in original). Lester Brown's Worldwatch Institute, founded in Washington, DC, in 1974, seized on these findings in white papers and journal articles, enlarging desertification into a worldwide emergency. In the pages of *Ambio*, Worldwatch's Erik Eckholm (1975, 138) wrote: "Usable land is being converted to waste by the influence of man and livestock. . . . Deserts are creeping outward in Africa, Asia, and Latin America. . . . Once lost to desert sands, land is reclaimed for human benefit only at enormous cost." Expanding cultivation of marginal land for crops, combined with growing livestock numbers, was "a sure-fire formula for overgrazing, wind erosion, and desertification" throughout the world's arid and semiarid lands (141). The tragedy of the commons lurked behind the problem: "The basic dilemma is that what is rational for, or even essential to, the survival of the society often flies in the face of what is rational for, or even essential to, the survival of the individual" (143). In both the USAID report and Worldwatch articles, desertification was understood as a human-caused disruption of the "delicate balance" or equilibrium of people, livestock, and the environment in arid and semiarid areas.

The severity of the Sahel drought peaked in 1972, when rainfall in some areas was zero, and the famine prompted the United Nations to create a special office dedicated to the Sudano-Sahelian region. The UN General Assembly passed a resolution in 1974 calling for an international conference on desertification, held in Nairobi, Kenya, in 1977, from which soon emerged the UN Plan of Action to Combat Desertification. The plan asserted that, "at least 35 per cent of the earth's land surface is now threatened by desertification, an area that represents places inhabited by 20 per cent of the world population. Each year 21 million hectares of once-productive soil are reduced by desertification to a level of zero or negative economic productivity, and six million hectares become total wasteland, beyond economic recoverability" (quoted in Thomas and Middleton 1994, 52).

The imagery of sand dunes marching across the landscape was spectacular, but its veracity did not stand up to scrutiny. Accurate data on African rangelands were notoriously limited (Grove 1974; Sandford 1976), and the figures in the UN plan were highly speculative. By the mid-1980s, scientists had refuted the claim that the Sahara was expanding, and many questioned whether desertification was even a coherent or meaningful concept. Follow-up assessments in 1984 and 1992 struggled to define desertification, let alone measure it, although the classic categories of range condition—excellent, good, fair, and poor—served as one basis for defining desertification classes (Thomas and Middleton 1994, 54). By the 1990s, some scholars openly denounced desertification as a myth (Thomas and Middleton 1994). But institutional developments had by then far outstripped scientific investigations, and in 1994 the UN Convention to Combat Desertification was adopted in Paris. It entered into force two years later and has been ratified by 195 nation-states and the European Union. Only one country, Canada, has denounced it, in 2013.

In hindsight, the most striking thing about desertification is that no comparably severe indictment had previously been leveled against ranchers in the United States, even though the ecological evidence for such a case had been present in the range science literature at least since the 1930s (Hutchinson 1996). But as Diana Davis (2007, 171) has stated, "The idea of desertification itself is in fact a colonial construction, a concept with little basis in empirical evidence initiated and propagated by those with a poor understanding of arid-land ecosystems." It had from its origins served the purpose of blaming Africans for the putative creation of desert conditions in the area known from antiquity as "the granary of Rome." According to this ideological construction, the culprits were nomadic pastoralists and their livestock, who were blamed—like the sheepherders of the late-nineteenth-century American West—not only for overgrazing but also for incendiarism. The development experts and neo-Malthusians unwittingly recapitulated this history, and the US Southwest stands as the exception that proves the rule: the basis of desertification was more political than ecological, and its target was specifically pastoralism, *not* ranching. As Cambridge geographer A. T. Grove (1974, 151) put it in 1974, "Perhaps the main problems are presented by pastoralists, whose traditional systems do not fit neatly into the framework of a modern state."

Ideas about rangelands played an outsized role in the politics of international development in the second half of the twentieth century, whether on or beyond rangelands, with or without the active participation of range scientists. Carrying capacity, desertification, and the tragedy of the commons alloyed apocalyptic neo-Malthusian fears of overpopulation to the scientific authority that range science had acquired between the two world wars, helping to legitimate draconian interventions not only in the lives and societies of pastoralists but in the reproductive practices of marginalized people around the world (Connelly 2008). The textbook definitions of range management presented the field as apolitical and of value to anyone interested in rangelands and their inhabitants anywhere.[3] But the contents of the textbooks—from vegetation measurements and utilization guidelines to brush control and improved breeding—were not merely tools and techniques; they also contained the blind spots and assumptions described in earlier chapters, which were themselves produced by the politics of US rangelands over the preceding century. It was those politics, in fact, that had compelled range science to *appear* so apolitical, empowering what James Ferguson (1990) famously called "the anti-politics machine" when deployed as the scientific authority behind Third World pastoral development.

This was hardly the only source of pastoral development's failures, embedded as it was in a much larger context of historical and contemporary power asymmetries (Baker 1984), but it played an important contributing role.

Most US range scientists paid little attention to rangelands overseas throughout this period, and their involvement in the rise of desertification as a global emergency appears to have been minimal. Their discipline was remarkably inward looking, even two decades after the explosion of interest in (and funding for) pastoral development. In 1976–1977, 88.5 percent of the citations in the *Journal of Range Management* were to US or Canadian sources—down from 95.8 percent in 1948–1951 but still highly unrepresentative of the world's rangelands (Vallentine 1979). The first International Rangeland Congress was not convened until 1978, in Denver; of the 560 attendees, 420 were from the United States and another 30 were from Canada; 25 came from Australia or New Zealand, while only 39 hailed from anywhere in Africa. Perhaps not coincidentally, 1978 was also when the term *desertification* first appeared in the *Journal of Range Management.* E. G. Van Voorthuizen, a range management advisor to USAID employed by the State Department, published a general appraisal of the phenomenon. Breaking with development dogma, he praised the traditional practices of nomadic herders such as the Maasai, and he pointed out that borders imposed by colonial powers and retained following independence had arbitrarily divided such groups' territories and disrupted their livestock management. He also challenged the tragedy of the commons thesis: "It has not been demonstrated, conventional and learned wisdom notwithstanding, that communal access to the range per se is in fact necessarily destructive over the long run" (Van Voorthuizen 1978, 380). But he did not question the premise that desertification was occurring, nor did he express any doubts about the relevance and value of range science. Instead, Van Voorthuizen complained that political and social obstacles had prevented development programs from implementing range management properly. Steps that were more socially acceptable—such as borehole wells and livestock disease control—had been implemented, whereas more controversial or expensive measures—such as stocking reductions or fencing—had not, and this had led to livestock increases that triggered desertification. Moreover, pastoral development projects had been implemented by teams of experts in which range managers were outnumbered by sociologists, economists, engineers, and others. "The range manager therefore should not be burdened with the blame for resource degradation resulting from social and political approaches to problems," he concluded (Van Voorthuizen 1978, 379–80). H. N. Le Houérou, a range expert at the International Livestock Center for Africa who had also worked for the FAO, reached similar conclusions in a more detailed assessment two years later. He argued that the social and political obstacles to effective range management—including communal tenure—would have to be removed "en bloc" as "a package deal," because "single measures are not operative" (Le Houérou 1980, 45). In other words, range science itself was innocent in the failures of pastoral development, and if properly applied it could still remedy problems in places like the Sahel. It would take researchers from outside of US range science—geographically and intellectually—to reveal its blind spots and formulate alternatives.

SEVEN

Till the Cows Come Home

Overseas Failures and Critiques of Range Science

In the three decades following World War II, while range scientists pursued artificial improvement of rangelands in the United States and development agencies endeavored to modernize range livestock production abroad, the core theory of range science remained basically unchanged. Criticisms of the Clementsian paradigm were lodged on specific points, such as the problem of accelerated soil erosion, which could make the return of climax conditions effectively impossible (Ellison 1949), and how it could be that secondary succession was sometimes more evident in conservatively grazed than ungrazed sites (Ellison 1960). How to measure grazing capacity was an ongoing subject of debate (Renner 1948, 1951; Stoddart 1953), and a handful of range scientists in the Southwest concluded by the 1960s that carrying capacities did not exist or could not be determined there (chapter 3). Empirical evidence at odds with successional theory continued to accumulate, such as the lack of correlation between grazing intensity and vegetation composition after twenty-three years of careful measurements at the Central Plains Experimental Range (Hyder et al. 1966; cf. Joyce 1993) and the inconsistency of results from a variety of areas protected from grazing for long periods (Gardner 1950; cf. Rice and Westoby 1978). Ecologists outside of range science actively debated successional theory, and a few challenged its applicability to arid ecosystems altogether (e.g., Gardner 1963), but no coherent alternative was forthcoming among range scientists. Meanwhile, the prospect of an artificial solution to degraded rangelands, however elusive, allowed agencies, ranchers, and scientists alike to treat the theory's main empirical failures as rectifiable anomalies rather than fundamental flaws. The third edition of the discipline's standard textbook, published in 1975, opened with the observation that, "In the more than 30 years since the appearance of the first edition of *Range Management*, there have

Table 7.1. Trends in rangeland condition on public domain (Bureau of Land Management) lands in the western United States: Percentage of land in three trend categories, 1930–1975

Forage condition	1930–1935	1955–1959	1975
Improving	1	24	19
Unchanged	6	57	65
Declining	93	19	16

Source: Hadley et al. 1977, 545.

been many changes. . . . Nevertheless, no new conceptual framework differentiates the field of range management now from then" (Stoddart, Smith, and Box 1975, ix).

Meanwhile, the political struggle over US public lands grazing grew ever more contentious, especially with the rise of environmentalism from the late 1960s onward, and successional theory was the taken-for-granted model underlying the views of scientists, agencies, ranchers, politicians, and environmentalists alike. How many livestock should be permitted to graze on the nation's public rangelands? Ecologically, economically, and politically, this was what all sides believed to be most important. The agencies gradually succeeded in reducing stocking rates despite ongoing resistance from livestock associations and many ranchers: as best as can be determined, actual stocking matched estimated carrying capacities on most Forest Service grazing allotments by the 1960s or 1970s (Alexander 1987a); permitted animal-unit-months on Bureau of Land Management lands declined by roughly half between 1950 and 1980 (Chambers et al. 2015). Broadly speaking, conditions by then had improved or stabilized relative to the open range period, but "original" or climax conditions had not been recovered. Fires continued to be rare and suppressed, shrub encroachment proceeded in some regions, and forage production swung back and forth with the highs and lows of rainfall, sometimes exceeding what stable numbers of livestock could consume and sometimes falling short. But with the possible exception of riparian areas and regionally significant invasive plant problems, western rangelands were neither degrading nor improving notably (table 7.1)—they were stuck, just like the antagonists in the debates about them.

Translated into the political realm, successional theory and the unquestioned obsession with stocking rates effectively guaranteed that the rangeland conflict would continue "till the cows come home." At the core of the impasse was the concept of carrying capacity itself. On the one hand, it was a static ideal fixed by climate and soils, to which any given piece of rangeland could and should return if managed properly—sometimes termed the "original capacity" of the range (Sayre 2010). On the other hand, as a measure of actual vegetation and forage conditions, it varied over time and could deviate significantly from the ideal for prolonged periods; it could also be increased by investments in revegetation or in stock trails or artificial water supplies that made more forage accessible to livestock. If a range was not in "excellent" condition—that is, at or near its climax—then its "actual capacity" was lower than its original capacity, and stocking should be set at this lower rate to allow recovery by secondary succession. Faced with a case of degraded range, agency officials could assure the public that restoration was possible by reducing stocking; environmentalists could demand complete destocking, convinced that this would bring about restoration most quickly; and ranchers could insist that conditions would improve "if and when it rains" (American National Live Stock Association 1938). All spoke the language of carrying capacity, but in effect they talked past one another.

Thus, when an alternative to the conventional model of rangeland science and management did finally coalesce, it came not from within US range science but from overseas, boomeranging back, as it were, after the western range had gone global. It emerged between 1967 and 1993 from at least three concurrent sources, initially separate but subsequently converging, overlapping, and borrowing from one another. A thorough analysis of all three far exceeds my scope here; some of the research was highly technical, and the social history is not well documented in published sources. Rather, this chapter presents an abbreviated review, simplifying many of the scientific details and focusing on how the emergence of an alternative was conditioned by the history recounted in preceding chapters—that is, by what it was an alternative *to*.

(1) Reacting to the failures of pastoral development projects, especially in Africa, European social scientists criticized the US ranching model as socially, economically, and environmentally inappropriate to pastoral societies and to drier and more variable rangelands. They did not directly challenge range science orthodoxy at first, but by 1990 they absorbed and amplified the findings of ecologists emerging from sources (2) and (3).

(2) Australian government agencies began to institutionalize scientific research on rangelands, which on that continent were disproportionately arid and semiarid. Although the Australian scientists drew heavily from US range science, especially at first, they soon recognized that Clementsian theory did not fit what they were observing in their own country.

(3) The International Biological Program (IBP), whose US component was well funded by Congress between 1967 and 1974, provided an institutional

context for rangelands research separate from the established, Forest Service–dominated system, and it constituted a network that connected ecologists from Africa, Europe, the Middle East, Australia, and the United States. The IBP's commitment to systems ecology and modeling meant that these scientists brought new kinds of tools and training to bear on range science questions and the findings of previous US range research and that they could work at larger spatial and temporal scales than was practicable for field-based methods.

Not surprisingly, formulating an alternative to the prevailing ideas of range science involved focusing on issues that had been obscured by the blind spots described in earlier chapters: exclusive land tenure and fencing, predators and rodents, fire and succession. The social scientists, for example, pointed out that measuring and enforcing carrying capacities presupposed fencing or some other means of containing herds within specific areas. That such control would give livestock producers an incentive to reduce their herds and invest in expensive breeds or other improvements further assumed exclusive land tenure, such that no one else's herds would have access to the same range. Finally, stocking rates fixed at a conservative percentage of average forage production failed to recognize the economic rationality of stocking more flexibly in highly variable arid and semiarid rangelands. Range scientists had never questioned the original need for fencing and leases to control stocking rates in the United States, and they had scarcely considered them at all since passage of the Taylor Grazing Act in 1934. Their discipline had assumed from its inception what pastoral development aspired to produce: a capitalist form of rangeland livestock production based on systems of land tenure, credit, property, and social relations that did not obtain in most Third World rangeland settings. Ranching, not pastoralism, was its unquestioned object. Studying pastoralists exposed these assumptions and opened up questions that range scientists had neglected or ignored.

Concurrently, the ecologists illuminated what had been obscured by Clementsian theory, predator control, and fire suppression, and they crossed paths—with one another and with US rangelands and range science—in the early 1970s through the IBP. The difficulties they encountered there in trying to model arid and semiarid ecosystems provoked them to develop another approach to understanding the dynamics of plants, herbivores, and rainfall. They applied predator-prey models from wildlife ecology and population biology to the relationship between herbivores and forage plants, for example, and they crafted a framework that could account for the empirical observations of range scientists and ecologists working in the southwestern United States, Australia, and Africa and that challenged the fundamental premises of Clementsian successional theory in range management.

By the late 1980s, a network of scientists connecting these three sources had formulated a "nonequilibrium" or "disequilibrium" approach to range science, based on "state and transition" models of vegetation dynamics rather than successional theory. Whether this alternative will enable resolution of the problems afflicting rangelands today remains to be seen. What is clear, however, is that the conditions that made its emergence possible were not merely scientific but also sociological and geographical: institutional developments in the United States and overseas, political and economic circumstances that conditioned those developments, and the disconnect between range science and range management that had widened over the preceding four decades.

Social Scientific Critiques of Pastoral Development

Numerous scholars have chronicled the almost universal and sometimes drastic failures of development projects that attempted to impose ranching and range science on pastoral areas around the world (Sandford 1983; Bennett 1984; Ferguson 1990; Gilles and de Haan 1994; Fratkin 1997; Forstater 2002; Markakis 2004). By the late 1980s, the World Bank and other agencies had "essentially given up on pastoralists and arid lands in Africa and [were] investing their resources elsewhere" (Ellis and Swift 1988, 451). My focus here is less on the programs and their flaws than on the responses that they provoked among scientists.

Three Swedish anthropologists were among the first to challenge the portrayal of pastoralists in development circles. In the same issue of the journal *Ambio* in which Eckholm proclaimed desertification a worldwide threat traceable to livestock grazing, Carl Widstrand (1975) insisted that pastoralists' tendency to build up their herds was in fact rational if understood in context, and that external forces were primarily responsible for any overstocking or range degradation that had recently occurred in Africa. "The pastoral livestock operation is *not a capitalistic undertaking* aimed at producing a marketable surplus. Its aims are rather to provide a good, regular supply of food for the family, to enable them to survive physically and socially and to maximize the chances of their surviving prolonged droughts and other risks" (Widstrand 1975, 149, emphasis in original). Pastoralists' use of land was opportunistic and flexible, keyed to the variability of rainfall and forage availability. Artificial water sources and veterinary medicine had encouraged herd growth, Widstrand noted, even as relatively fertile areas had been converted to crop production and mobility had been curtailed, reducing forage supplies. Gudrun Dahl and Anders Hjort published detailed book-length arguments along similar lines shortly thereafter (Dahl and Hjort 1976; Dahl 1979).

Two British economists pushed this line of critique further by emphasizing the irrationality of fixed carrying capacities and stocking rates in highly variable environments. Stephen Sandford, an agricultural economist who grew up in colonial Tanganyika (now Tanzania) and worked for British development agencies in Ethiopia, Botswana, and elsewhere, questioned the logical coherence and empirical support for prevailing theories of degradation in pastoral areas: "For half a century or more there has been near-universal agreement that conditions in the rangelands are deteriorating. . . . Firm evidence of changes in these conditions is scanty" (Sandford 1976, 45–46). "Scarcely anybody," he opined, "comes to consider the problems of rangelands or pastoralists with a completely open mind. Whether they be academics or officials of international agencies or of governments of developing countries, they will have absorbed, even if unconsciously, something of these theories, and will be predisposed to analyse a particular situation in a certain way and to choose particular ways of putting things right" (47). Sandford insisted that, "Clearly it is not the level of *average* rainfall in an area that causes deterioration or even absolute poverty. Rather it is the variation over time, or a declining trend, that brings problems" (52).

Writing about the Sahelian disaster, development economist Jeremy Swift, who worked on conservation in Africa for the Food and Agriculture Organization of the United Nations in the late 1960s, noted, "Plant production can, within 'normal' rainfall variation, double or halve from year to year, and in exceptional years like 1972 can drop to zero. Animal production varies in proportion." Stocking rates fixed to match dry years "would leave unexploited huge pasture areas in good years," whereas rates set for wet years would be disastrous in bad ones (Swift 1977b, 459). Like Widstrand, Swift pointed to exogenous political and economic forces as the causes of range degradation in the Sahel:

> The subsistence pastoral economy has as its aim survival, not profit, and its objectives are not at all compatible with those of a market livestock economy; the beginning of a transition from one to another has reduced the ability of pastoralists to exploit a difficult environment with an adequate safety margin. . . . For a number of years they did well out of the increased security, better communications and technical improvements such as the new wells. But they were losing ecological and economic flexibility; they were being forced into a

> market economy and their own self-help mechanisms were breaking down. The natural environment was being used more intensively, in a less ordered and controlled manner, allowing little margin for variations in rainfall or the pasture's limited capacity for self-renewal. Increased desertification and increased susceptibility to famine were the consequences of these long-term trends in the Sahel. (Swift 1977a, 175–76)

That fixed stocking rates at average carrying capacities meant not utilizing large volumes of forage in wetter years had been noted by US range scientists since Clements, but it had been viewed (by administrators at least) not as inefficient or wasteful but as a wise investment to offset the effects of heavy grazing in drier years. Sandford pointed out that fixing such a rate was not merely an ecological matter, however, but also a question of the probability of droughts of various severities: one in ten years, one in fifty years, and so on (as hydrologists and engineers measured floods). He argued that the losses avoided during drought by always stocking at low rates should be weighed against the opportunity costs of understocking in wetter years (Sandford 1976, 52). In 1982, he published a simple modeling experiment that quantified this approach by comparing two scenarios: the conventional strategy of range science—which "maintains a population of grazing animals at a relatively constant level" and which he labeled *conservative*—and an *opportunistic* strategy characteristic of traditional pastoralists, which "varies the number of livestock in accordance with the current availability of forage" (Sandford 1982, 62). The model showed that the opportunity costs of production foregone under a conservative strategy increased sharply with increasing variability: stocking to match conditions during a once-in-twenty-year drought would reduce overall output 33 percent if the coefficient of variation of rainfall was 20 percent, 58 percent if it was 35 percent, and 82 percent if it was 50 percent (Sandford 1982, 69). He concluded that, "For the present purpose an estimate of the coefficient of variation of the annual rainfall of an area is more important than an estimate of its average value" (66) and that "pastoralists' frequent reluctance to embrace a conservative strategy may be better founded than non-pastoralists in the past have realized" (72).

In 1983, Sandford published an influential, comprehensive review and analysis entitled *Management of Pastoral Development in the Third World*. Drawing on the work of anthropologists, ecologists, economists, development agencies, and range scientists, Sandford (1983, 15) challenged the "overwhelming concern with control over and stabilization of livestock numbers" in what he termed the "mainstream view" of pastoral development. He observed that "One major problem has been an obsession with the doctrine of the 'Tragedy of the Commons' and a misplaced faith that the private ownership of land is the only answer to the dilemma that the doctrine poses" (16). And he offered a list of ways in which the ranching model did not make sense in most pastoral contexts:

> The American and Australian models are particularly unsuitable for most developing countries, originating as they did in peculiar historical settings where the interests of the previous inhabitants of pastoral areas were not taken into account, where the species of domestic livestock on which pastoral development focused did not previously exist on a significant scale if at all, where the general economy as a whole was characterized by labour shortage rather than by surplus, and where a large and wealthy non-pastoral sector could be called on from time to time to provide the resources with which to rebuild a pastoral sector suffering from collapse. (6)

The social scientists' critiques, and especially those of Sandford,[1] caused major development agencies such as the World Bank to reconsider and curtail their project plans for pastoral areas in the 1980s (Fratkin 1997), but the response among US range scientists was limited. Although some of the critiques challenged the idea of ecological equilibrium (e.g., Swift 1977a), none identified Clements or successional theory as an underlying flaw, so the problems overseas could still be attributed to the social and political circumstances of the developing world rather than to range science itself. In a favorable review of Sandford's book in the *Journal of Range Management* in 1984, for example, Charles Poulton recognized that the critique had the potential to force a far-reaching reevaluation of the discipline. "His assessment of the consequences of the 'mainstream view' is enlightening and will challenge many to reassess their position. It may, in fact, open the minds of some pastoral technicians from the developed world to a new and different attitude. . . . The book needs to be studied and contemplated in depth." But Poulton tacitly indemnified the ideas and practices of range science by emphasizing the misapplication of "*developed world technology*, management systems and practices" that had been "interjected into the pastoral situation without questioning need or appropriateness—merely because it may have worked reasonably well in a high cash-flow, economically developed society" (Poulton 1984, 478, emphasis in original). This argument could not be made for the case of Australia, however, where the socioeconomic conditions surrounding range livestock production were basically identical to those in the United States.

The Rise of Rangeland Science in Australia

The US model of leasing publicly owned rangelands to private producers for fees tied to carrying capacities had its historical origins in Australia, as noted before. States, rather than the federal government, administered most rangelands there, and sheep were more prominent (relative to cattle) than in the United States. A much larger proportion of Australia's rangelands were arid and semiarid, which made the research done in the southwestern United States particularly relevant. (Also, Australians use the terms *pastoralism* for what Americans call ranching and *station* for ranch; I will use the American terms to avoid confusion.) But as systems of production and management, the rangelands of the two countries were very similar, especially compared to pastoralism in the developing world.

Where the two countries differed was in how and when scientific research on rangelands found institutional form. Australia did not have a land-grant university system, which in the United States after 1862 had fostered colleges of agriculture in public universities, where range science found an academic home throughout the western states by midcentury. In Australia, scientific research on rangelands was very limited until the 1950s. The focus at first was on arid lands and deserts rather than rangelands, per se, prompted in part by UNESCO's Advisory Council on Arid Zone Research, launched in 1951. When the Commonwealth Scientific and Industrial Research Organization (CSIRO) created its first Arid Zone Research Liaison Officer in 1954, half of the fourteen existing research stations concerned with rangelands were less than a decade old. "They were small, scattered and poorly coordinated, and typically 'limited by a shortage of funds and difficulty in obtaining suitable scientific staff.' . . . Nearly half of the 'scientific research institutes' were demonstration sheep stations and experimental farms modestly funded by state agencies. . . . None of the Australian arid-zone research stations had the substantial support that those managed by the United States Forest Service enjoyed" (Robin 2007, 113). The first CSIRO research unit dedicated expressly to rangeland science was not formed until the late 1960s, at Alice Springs (Joss, Lynch, and Williams 1986, xii), and the Australian Rangeland Society was only founded in 1975, a generation after its US counterpart.

In the 1960s and 1970s, Australian rangeland research drew heavily on ideas and experts from elsewhere, and especially from the United States. Harold Heady from Berkeley and Thadis Box from Utah State were particularly influential (Joss, Lynch, and Williams 1986, xii). "Perhaps it was easier to take advice from the American south-west, India or Israel than from Canberra. From 1960 international arid-zone experts were the focus of conferences. By the 1970s, doctoral students were undertaking Australian research for degrees from overseas universities in arid-zone science. These included Utah, Tucson, Israel, Jodhpur and Guelph" (Robin 2007, 114). After 1978, CSIRO's *Arid Zone Newsletter* began to focus more narrowly on rangeland studies, as other arid lands ecologists found more specialized journals in other fields for their work. "The focus on 'rangelands' redefined the Australian desert as a hub of pastoral industry, in lands beyond agriculture and forestry. Rangelands gave an Australian place an international (particularly North American) professional context, and provided scientists working in Australia with an international audience for their publications" (Robin 2007, 118).

In conducting their own research, however, the Australian range scientists appear to have been less beholden to inherited ideas than their counterparts in the United States. In ways reminiscent of the earliest reports of the Division of Agrostology, the Australians were prepared—or compelled by circumstance—to examine all the possible factors that had influenced the landscapes they studied before fitting them into any preconceived theory. At the Alice Springs laboratory, for example, CSIRO scientists readily acknowledged the importance of rodents and fire suppression as well as livestock grazing, and they conducted experiments on burning (Griffin, Price, and Portlock 1983; Griffin and Friedel 1984, 1985). Moreover, they interpreted these impacts and their interactions as discontinuous, potentially irreversible, and keyed to weather events such as extreme drought or unusually heavy rainfall (Griffin and Friedel 1985).

There is not space here to review Australian range science research in detail, and in other lines of inquiry it was consistent with the US model (e.g., mechanical and chemical control of brush). As will be evident in the next section, its importance for present purposes was in providing an institutional context that could support scientists who were squarely focused on rangeland ecology and close enough to range science to digest its theoretical underpinnings yet sufficiently autonomous to question and challenge those underpinnings. It is telling, for example, that when Australia hosted the Second International Rangeland Congress in Adelaide in 1985, the first (and longest) panel concerned "Dynamics of Range Ecosystems," and it included several papers that directly disputed the Clementsian model (Joss, Lynch, and Williams 1986). In his closing address to the Congress, Thadis Box remarked that, "Here in this conference, the Clementsian concept of succession has been repeatedly attacked" (Joss, Lynch, and Williams 1986, 614). "Most American range workers," he went on, "have made their peace with this conflict. It is very old. . . . Practical range people have found the concept works well on the tall-grass prairie where Clements developed his theories; less well in other grasslands, and breaks down almost completely in the mountains and deserts." But he immediately added, "Attacked though it was, no new theories came out of this Congress to replace the Clementsian model. None were presented that can be tested and no new paradigm was proposed for our science of range management" (Joss, Lynch, and Williams 1986, 615).

Modeling and the International Biological Program

It would not be long before such an alternative was presented, based on a growing body of ecological research on rangelands in various parts of the world. It reflected broader shifts in ecology associated with the work of C. S. Holling (1973), Robert May (1977), and John Wiens (1977, 1984), who challenged the concept of equilibrium and specifically the notion that every ecosystem has a single stable state or climax to which it returns following disturbance thanks to feedbacks among its component parts. Empirical data and modeling were both suggesting, instead, that some systems had multiple stable states and that external forces or stochastic events might be as important as (or more so than) internal feedbacks. These ideas were not yet attracting the attention of many US range scientists, but they were of great interest to ecologists, and because drier, more variable ecosystems appeared particularly likely to display these dynamics, rangelands attracted their attention in new ways.

The International Biological Program provided the institutional context for the emergence of nonequilibrium rangeland ecology, albeit somewhat by accident. Inspired by the International Geophysical Year of 1957–1958, the IBP was proposed by the International Council of Scientific Unions in 1961 and ran from 1964 to 1974 under the banner of "the biological basis of productivity and human welfare." Some 150 scientific academies participated; like the Geophysical Year, the IBP envisioned an exhaustive collection of standardized data worldwide—in this case, on basic biological phenomena and processes. In the United States, it was supported by the National Science Foundation beginning in 1967 and by special congressional authorization after 1970, representing the first attempt to apply the "big science" model to biology (Kwa 1993; Aronova, Baker, and Oreskes 2010).

At the core of the US IBP was systems ecology, heavily influenced by cybernetic theory and engineering and inspired by the prospect of using simulation models to achieve control over nature. Neo-Malthusianism was at its peak in this period, as we have seen, and for lobbying purposes at least, the IBP married the agenda laid out by Vogt and Osborn twenty years earlier to the perceived power of computers (Kwa 1987). As Frederick E. Smith, director of the analysis of ecosystems for the program, wrote in the *Proceedings of the National Academy of Sciences* in 1968, "We have new and powerful tools in systems analysis and in high-speed computers. With such tools we have come to expect that the complexity of ecosystems will yield to analysis" (Smith 1968, 8). Smith envisioned systems ecology having "profound effects upon decisions to be made as man completes his domination of the earth" (5), and he helped persuade Congress to fund the IBP as a way to solve both domestic pollution problems and the world's food and population crises (Kwa 1987). The IBP was "a program to assess the capacity of this planet as a life-support system. By mapping and characterizing all the environments, both terrestrial and marine, and by making suitable measures of the present and potential production, the IBP hopes to obtain some practical measure of the earth's capacity to support mankind" (Smith 1968, 9). The challenge was similar to that of range reconnaissance fifty years earlier but orders of magnitude more complex: the human carrying capacity of the entire world.

The IBP divided the country into five biome-based program areas, and the first to be established was the Grasslands Biome program at Colorado State University, directed by George Van Dyne. Van Dyne had trained in animal husbandry and nutrition, and he held a joint appointment in the Departments of Range Science and of Fish and Wildlife Biology. He considered himself a range scientist and published regularly in the *Journal of Range Management*. But he had become engrossed in systems ecology and the new field of computer-based simulation modeling while working at Oak Ridge National Laboratory before coming to Colorado State. As head of the Grasslands Biome program from 1967 to 1974, he would oversee hundreds of biologists, mathematicians, engineers, and computer programmers, many of them far removed from established range science. Also concerned with rangelands was the Desert Biome program, established in 1969 at Utah State University.[2]

At its conclusion in 1974, the IBP was widely considered a failure, and Van Dyne had been forced out of his leadership of the Grasslands Biome program (Kwa 1993). Despite having amassed vast amounts of data from rangelands throughout the western United States (including the Jornada Experimental Range) and having built a huge (for its time) computer model of grassland ecosystems—called ELM, for Ecosystem Level Model, with 120 state variables and more than 1,000 parameters (Kwa 1993, 141)—the Grasslands and Desert programs had produced few practical insights for range management, and most ecologists disputed the value of the IBP for their discipline as well (Kwa 1993; Aronova, Baker, and Oreskes 2010). "The grandiose ideal of achieving total control over ecosystems, which around 1966 appealed so much to systems ecologists as well as Congressmen, was dismissed as 'hyperbole,' ten years later in a report in *Science*" (Kwa 1993, 155).[3] In retrospect, however, the IBP's failures were productive ones. In addition to generating vast data sets and hundreds of publications, it had pushed and revealed the limits of simulation models to capture the complexity of ecosystems, and this catalyzed valuable reactions among a handful of the hundreds of scientists whose paths had intersected through the IBP's international network. They were ecologists who combined an interest in rangelands with expertise in modeling, and they were not conventionally trained in range science. The five most central for present purposes were Mark Westoby, Imanuel Noy-Meir, Brian Walker, James Ellis, and David Swift.

Westoby went to Utah State University from Scotland in 1970 to earn his doctorate in wildlife ecology and to work with David Goodall doing modeling for the Desert Biome program. He wanted to apply predator–prey models to the relationship between herbivores and plants, and Goodall was one of the few people in the world who could advise such a project. The ELM model at the Grasslands Biome was considered the most advanced of its kind, according to Westoby, and the strategy it reflected was that "everything would go in the model . . . and a comprehensive understanding would emerge."[4] But as he attempted to build models for the Desert Biome, he concluded that some variables were unimportant, whereas other, important variables were being excluded for lack of field data about them. It was an endless regress: They could model the growth of annual plants as a function of rainfall, for example, but what if there were no annual plant seeds in the seed bank? They could model seed production, but harvester ants might consume them all—this, too, could be modeled in terms of ant movement as a function of soil surface temperature, perhaps, but then one needed data on the rate of seed burial, and so on. "To cut a long story short, the models weren't stable at all—the seeds went to zero." Since there was no way to collect all the potentially relevant data, including everything could not be the strategy. This led directly to Westoby's idea of the "pulse-reserve model" of arid zone ecology, in which rainfall events drove ecosystem dynamics: a pulse of water would trigger a ramifying burst of biological activity, including the production or replenishment of reserves to survive the subsequent dry period, when growth would again drop to zero. Westoby completed his

PhD in 1973 and took a position at Macquarie University in Australia two years later.

Noy-Meir spent a sabbatical year at the Desert Biome program in 1972. He had grown up in Argentina and Israel before earning a PhD at the Australian National University in Canberra, and he taught at the Hebrew University in Jerusalem from 1970 to 2007 (Seligman, van der Maarel, and Díaz 2011). While in the United States, he heard Westoby present the pulse-reserve model at an IBP meeting, and he used it to organize a set of review articles on arid ecosystems for *Annual Review of Ecology and Systematics* that he had come to Utah State to write (Noy-Meir 1973, 30). He proposed constructing "simulation models of deserts in terms of discrete events and qualitative states rather than continuous processes and variables," organized around "rainfall variation at several time-scales" ranging from years to within-days (Noy-Meir 1973, 30–31). The models suggested that strong biotic couplings with organisms or populations would not form in a pulse-driven ecosystem: when resources were available they were superabundant, so there was no significant competition for them, and the rest of the time the system was effectively dormant. As Noy-Meir (1979, 4) put it later: "In deserts, the dynamics of each population are determined mainly by its independent reaction to the environment, in particular to the highly intermittent availability of water. Desert plants and animals have no significant feedback effects on the important environmental factors, nor on other populations. Interactions between species are weak and their effects are negligible in comparison with the overwhelming effect of the weather." Noy-Meir's reviews "drove a shift away from equilibrium towards event-driven thinking for arid ecosystems" (Walker and Westoby 2011, 17). Back at Hebrew University two years later, Noy-Meir used graphs adapted from studies of predator–prey interactions to examine the stability of grazing systems under various rates and forms of plant growth and herbivory. These models provided a framework capable of accounting for multiple stable states of vegetation and for discontinuously stable systems characterized by tipping points between states (Noy-Meir 1975).

Walker grew up in British Rhodesia, studied rangelands in college at the University of Natal, South Africa, and then earned a PhD in plant ecology at the University of Saskatchewan. His focus was modeling, and while he was in Canada he met George Van Dyne, who came to speak about the IBP. Walker encountered successional theory at Saskatchewan in the person of his advisor, Robert Copeland, who had been a student of Clements's most famous protégé and colleague, John Weaver. But Walker also watched Copeland spar with a colleague, Ralph Dix, a botanist and critic of succession, and he visited the Jornada Experimental Range and noticed there that the Clements–Dyksterhuis framework was one that "people couldn't seem to challenge."[5] After his PhD, Walker taught in Rhodesia and South Africa, and he found successional theory difficult to square with his observations of savannas in Africa. Seeking an alternative, he arranged to spend part of a sabbatical year with Noy-Meir (and the rest with C. S. Holling) in 1978–1979. One outcome was an influential article in the *Journal of Ecology* (Walker et al. 1981) that used Noy-Meir's graphical analysis techniques to examine grass–shrub dynamics in grazed savannas. Walker moved to Australia in 1985 to head the Division of Wildlife for CSIRO.

At about the time of Walker's sabbatical, Noy-Meir invited Westoby to contribute to a special issue of the *Israel Journal of Botany* that he was editing, resulting in "Elements of a Theory of Vegetation Dynamics in Arid Rangelands," which confronted range management and succession directly. "Insofar as range management has a theoretical basis, it is the concept of range succession," it began. "This concept has imperfections from two points of view. First, it has not changed to reflect recent changes in our understanding of successions generally, and of the processes that drive them. Second, there are a number of important range types, particularly in climates where rainfall is limiting, whose dynamics cannot be interpreted with the classical concepts of range succession" (Westoby 1979, 169). In presenting his alternative theory, Westoby cited research from around the world, including studies from the Santa Rita Experimental Range that demonstrated the irreversibility of shrub encroachment, the multiplicity of persistent vegetation states, and the importance of seemingly random events (or sequences of events) such as drought, acute overgrazing, heavy summer or winter rainfall, and fire (or lack thereof). His theory rested much less on species composition—the heart of successional approaches—than on process-based interactions keyed to different "types of time," defined in relation to pulses of rainfall, and "different permutations of drought and grazing sequences" (Westoby 1979, 187). Episodes of extremely wet or dry weather could have effects on vegetation that lasted for decades, and livestock grazing could be helpful or harmful in ways that depended less on stocking rates than on its timing. His conclusion upended the assumptions that had reigned in range science since Sampson's famous 1919 bulletin:

> The classical model of range succession thinks of vegetation, at any given stocking rate, as in a stable equilibrium, held in place by the balance of two steadily-applied opposing forces, grazing pressure and the tendency of the

> vegetation to change towards climax. Correspondingly, range management has set as its first task to determine the equilibrium condition, and hence the stocking rate, appropriate to each area of rangeland. However the forces of intrinsic change and of grazing pressure in arid vegetation do not operate steadily, and not necessarily in opposing directions. Changes in one direction often cannot be reversed by changing the pressure. Given grazing pressures may have no significant impact on vegetation much of the time, but in conjunction with a particular weather event may quickly produce radical changes. Range management has handled these problems by continuing to set equilibrium stocking rates, but making them conservative. Aside from using vegetation inefficiently, this policy can cause range problems of its own; it allows stock to select the most palatable species, so that mean palatability of the vegetation decreases. Basically an equilibrium approach to stocking is ill-adapted to arid vegetation. Changes in the vegetation are often responses to exceptional events, rather than to average conditions. (Westoby 1979, 190–91)

When Noy-Meir and Walker went to Australia in 1984 to present a paper at the Second International Rangeland Congress, Noy-Meir introduced Walker to Westoby, and the three of them decided to write a paper together. According to Westoby, by this time, "we felt that the academic objections to Clementsian theory had already been amply discussed, and the point was to develop some sort of alternative that would be workable for managers."[6] Published in 1989 in the *Journal of Range Management*, "Opportunistic Management for Rangelands Not at Equilibrium" has since succeeded Sampson (1919) and Dyksterhuis (1949) as the most influential paper in US range science. The authors presented the "range succession model" and the "state and transition model" as "two ends of a spectrum of possibilities" (Westoby, Walker, and Noy-Meir 1989, 266), but they suggested that the state and transition model is probably superior "in many rangelands, in particular in arid and semiarid regions" (268). In such places, "Vegetation changes in response to grazing have often been found to be not continuous, not reversible, or not consistent" (268). Using three sites from Australia and South Africa as examples, they represented rangeland vegetation dynamics with boxes-and-arrows diagrams, each box representing a stable state that differs from others in some way important to management, and arrows symbolizing transitions between states whose causes can be described or explored. "Research would aim to construct the catalogue of possible states of any given rangeland" and to conduct "experimental tests of hypotheses about the various transitions" (271).

At the center of their argument was the need for scientists, managers, and administrators alike to be opportunistic. "Many of these transitions can only occur given an appropriate climatic sequence, plus the hypothesized management with respect to grazing, fire, seeding, etc. Experiments on such transitions would be planned on a contingency basis. They would be put into operation not on a fixed research schedule but when the relevant climatic conditions arose. . . . [Otherwise,] by the time a proposal had been written and money allocated from the next funding cycle, the opportunity would have passed. This is why few such opportunistic experiments have been done." Range managers, similarly, "would see themselves facing an oncoming stream of events, a mixture of opportunities and hazards. Their objective would be to seize the opportunities and evade the hazards, so far as possible" (Westoby, Walker, and Noy-Meir 1989, 271). Government agencies, meanwhile, would "Drop the assumption that inaction or conservative grazing is safe. In many situations moderate grazing leads to range deterioration. Sometimes very heavy grazing is a constructive thing to do. Often, burning is a constructive thing to do. Legislation and regulations need to free managers to intervene positively. Where possible, regulation should focus not on stocking rates but on changes in the actual state of the land or the vegetation" (273).

Range Ecology at Disequilibrium

At the end of the 1980s, the efforts of the three groups described above converged in a single volume entitled *Range Ecology at Disequilibrium: New Models of Natural Variability and Pastoral Adaptation in African Savannas* (Behnke, Scoones, and Kerven 1993). It grew out of a conference convened at Woburn, England, in 1990 by the Overseas Development Institute, a think tank supported by the British government, where Stephen Sandford had previously worked as the first head the Pastoral Network, an information exchange with about 600 members from around the world. Most of the conference attendees were ecologists working in pastoral areas in Africa, based at universities in Europe, Africa, Australia, and the United States; most of the US participants were affiliated with range science programs or with Colorado State University's Natural Resources Ecology Laboratory, which had been created to support the IBP's Grasslands Biome program and had continued thereafter. Both Sandford and Westoby were invited but were unable to attend.

The Woburn conference was inspired by the work of James Ellis and David Swift, two US ecologists who had been at the Natural Resources Ecology Laboratory since IBP days, originally recruited to assist with modeling. Since then, they had developed a major long-term interdisciplinary study of the

Turkana, a pastoral society in northern Kenya, combining detailed ethnographic and ecological research in close collaboration with anthropologists and local pastoralists. In an invited article in the *Journal of Range Management* in 1988, Ellis and Swift brought together the arguments of Sandford and Jeremy Swift about opportunistic grazing strategies, the insights of Noy-Meir and other ecologists about abiotic drivers in arid and semiarid systems, and nine years of their own team's work on the Turkana. "The near universal failure of pastoral development," they wrote, cannot be attributed merely to technical mistakes or local resistance but rather "suggests that something more fundamental is amiss" (Ellis and Swift 1988, 451). That fundamental something, they argued, was the "equilibrial paradigm" of range science itself:

> The paradigm assumes that the ecosystems occupied by pastoralists generally function as equilibrial systems which are regulated by density-dependent feed-back controls; however, pastoralists override these feed-back controls to the detriment of themselves and their ecosystems. If this assumption is accepted, it is logical to reason that internal alterations in system structure can correct the imbalances and restore the system to equilibrial conditions. The most obvious adjustments to make are those involving the number of livestock per unit area. Hence two types of development procedures follow: reduction of stocking rates and alteration of land-tenure systems. (452)

Their own data, by contrast, pointed to the overriding importance of "external control mechanisms, i.e., drivers, which are not subject to feed-back control from within the system" (Ellis and Swift 1988, 452). Annual dry seasons reduced forage production to near-zero levels, and the short rainy season sometimes failed for one or more years, resulting in severe drought. Livestock herds declined during multiyear droughts by 50 to 70 percent, and when good rains returned, the herds were too small to have a major impact on the resulting burst of plant growth. "Thus, while livestock may, in the long-run, alter the structure and composition of the plant community, they appear to have no role in regulating yearly plant production, only a minor role in regulating biomass levels and consequently little or no role in regulating the amount of forage available. The strong force exerted by climate on forage production and the minimal influence of livestock on forage availability means that there is little opportunity for the development of strong feedbacks from livestock to plants" (455).

In a nonequilibrial setting, the conventional range science model was profoundly flawed. "The ecosystem is not balanced, does not operate in an equilibrial fashion, and cannot be treated as if it did" (Ellis and Swift 1988, 457). Rather, "ecosystem dynamics are dominated by the stochastic perturbations of multi-year droughts. Under these conditions large-scale destocking would result in immediate deprivation for pastoralists even during mild stress periods. Likewise, confining pastoralists to grazing blocks or ranches would reduce the spatial scale of exploitation and result in disaster during serious droughts. The obvious conclusion is that conventional development procedures are destabilizing influences in ecosystems which are dominated by stochastic abiotic perturbations and which operate essentially as non-equilibrial ecosystems" (458). The Turkana themselves understood the need to expand the spatial scale of their activities during drought, whether to sell or graze their livestock or to disperse themselves outward into non-pastoral livelihoods until the rains returned. "An extensive spatial scale is a prerequisite for a successful pastoral system where droughts are frequent. Reductions in scale or confining pastoralists to ranches is an invitation to disaster" (458).

Ellis and Swift attended the Woburn conference and wrote the synthetic chapter of the resulting book. Other scholars presented cases comparable to the Turkana from Zimbabwe, Ethiopia, South Africa, Mali, and the Sahel. Only the case with the highest rainfall, in the Ethiopian highlands, seemed to fit the equilibrium model; the one from Zimbabwe straddled the line, showing signs of equilibrium behavior during wetter years and nonequilibrium behavior during droughts. Two other chapters scrutinized the concept of carrying capacity. In their introduction, Roy Behnke and his counterpart from the International Institute for Environment and Development, Ian Scoones, announced that "What were once anomalous individual field cases are now increasingly linked into an internally consistent, alternative theory of the functioning of savanna and rangelands . . . the mainstream view of range science is fundamentally flawed in its application to certain rangeland ecologies and forms of pastoral production. If range management is to be of any use in these settings, conventional theories and recommended management practices require not minor adjustment but a thorough re-examination" (Behnke, Scoones, and Kerven 1993, 1–2). Drawing on work in wildlife population biology, they pointed out that the optimum economic carrying capacity, to which commercial ranching aspired, was much lower than the ecological carrying capacity, appropriate for pastoralism. In the final chapter, two Australian range scientists, Mark Stafford Smith and Geoff Pickup, drew parallels between vegetation-fire-grazing dynamics in African and Australian settings, highlighting the importance of processes at multiple spatial and temporal scales. "The majority of habitats do not fit the Clementsian model," they concluded, "and the model has particularly severe limitations when applied to episodically-driven arid and semi-arid rangelands" (Behnke, Scoones, and Kerven 1993, 218).

In light of the tenacious grip that Clementsian theory had held on range science since the 1930s, the nonequilibrium alternative found remarkably rapid acceptance in the closing decade of the twentieth century. There were no pitched debates in the pages of the *Journal of Range Management*—indeed, there were no debates at all, as the core ideas and relevance of the new theory found few, if any, detractors. Rather, range scientists in the United States and elsewhere began reinterpreting previous research findings in terms of thresholds and multiple stable states while directly challenging the successional assumptions of inherited concepts and practices (Friedel 1991; Laycock 1991; Svejcar and Brown 1991). In both methods and subjects, the nonequilibrium approach allowed the scope of range science research to expand to include, for example, simulation modeling and the roles of fire and climatic variability in vegetation change (e.g., Fuhlendorf, Smeins, and Grant 1996; Fuhlendorf, Briske, and Smeins 2001). In 1994, a committee of the National Academy of Sciences candidly admitted the flaws of Clementsian theory as it had been adopted by federal government agencies; three years later, the Natural Resources Conservation Service (successor to the Soil Conservation Service) began revising its catalog of "range sites" (the basis of range condition assessments since the days of Dyksterhuis and Parker) under the rubric of "ecological sites" as a step toward integrating state-and-transition models into its *National Range and Pasture Handbook*. Consistent with the history recounted in earlier chapters, scientists at the Jornada Experimental Range are at the forefront of this work (Bestelmeyer et al. 2003), and the Bureau of Land Management is also actively involved, whereas the Forest Service appears to be more resistant to overhauling its system.

What accounts for this rapid acceptance? The most obvious answer is that the inadequacies of Clementsian theory were already widely known. As Thadis Box had said at the close of the Second International Rangeland Congress in Australia in 1985, "practical range men" had long since resigned themselves to the limited utility of successional theory outside of the Great Plains. For many of them, the lack of an alternative theory may have been of little concern, at least for "practical" purposes, and when one did arise, they greeted it as obvious or of minor consequence. But other, extrascientific circumstances also helped erode the erstwhile necessity of Clementsianism. The Forest Service exercised much less influence than before over range research as a whole, as evidenced by its relinquishment of the Jornada and Santa Rita Experimental Ranges in 1953 and 1988, respectively. It had largely abandoned range monitoring, as described in chapter 4, and it no longer dominated the job market for trained range professionals. Overall levels of livestock grazing on both Forest Service and Bureau of Land Management lands, as measured in animal-unit-months, had been in steady decline for decades, and the market value of many ranches had come to depend less on their official livestock carrying capacities than on the development potential and "amenity values" of their land (Torell et al. 2005). Always subordinate to forestry and fire suppression within the Forest Service, rangeland issues had by this time been surpassed by recreation and endangered species conservation as well; in some parts of the country, in particular the Southwest, range administrators were more preoccupied by legal battles with environmental groups than the latest debates in range science. For bureaucrats and ranchers as well as scientists, it no longer seemed plausible to believe, or worthwhile to pretend, that stocking rates were the key factor affecting western rangelands.

Acceptance is no guarantee that the new theory will succeed where Clementsian range science failed, however. The implications of the nonequilibrium model are far reaching, and many obstacles remain to be surmounted. Westoby, Walker, and Noy-Meir (1989) were correct to predict that engaging rangelands opportunistically would not be easy for scientists, managers, or administrators. How can a scientist or a funding agency invest in planning an experiment that might not happen until the right sequence of weather events happens to occur, years or even decades later? How can thresholds of change between states be predicted if they result from semirandom combinations of events and can only be definitively identified after the fact? For managers, the need to adjust rapidly in response to fickle and unpredictable weather has long been a challenge, and it is far from obvious how the new theory makes it any easier. Finally, for agencies and policy makers, crafting regulations that deemphasize stocking rates in favor of broader ecological outcomes feels like a leap of faith: who can be held responsible for those outcomes if they ultimately depend on the weather? How even to judge an outcome in the absence of climax as the bedrock standard of what a given piece of rangeland can and should be? The Forest Service and other federal land management agencies have endorsed "adaptive ecosystem management" in principle, but in practice such opportunistic flexibility runs counter to deeply entrenched legal-bureaucratic norms and may often collide with budgetary constraints as well.

For most everyone involved, the clear lines that organized judgments about rangelands in the past—between natural and artificial improvement, native and nonnative, excellent-good-fair-poor—have blurred. To take just one example: Under the successional model, a site dominated by Lehmann lovegrass was necessarily judged to be in need of improvement because Lehmann lovegrass is a nonnative plant in the United States and therefore cannot be part of the climax plant community. This was true wherever the site might be, now and forever. Under the nonequilibrium model, by contrast, things are more complicated. What is possible for a site can no longer be deduced from a static description of its "original" vegetation, and what is desirable now depends on what people value for that site and how much money is available to try to change it. All evidence indicates that Lehmann lovegrass is here to stay: it won't go away simply by removing livestock, nor even if fire is applied. It may be less desirable than native perennial grasses but still preferable to the mesquite-dominated vegetation it replaced. The current condition might be considered "excellent" in a place where watershed function is the priority, "good" in a place dedicated to livestock grazing, and "poor" where biodiversity concerns are greatest. Whether "improvement" is possible may depend not only on unpredictable weather and site-specific factors but on the dispositions and resources of individual people. In short, history and geography matter in ways they didn't under the equilibrium model, and policies can no longer be crafted in abstraction from those considerations. Not only must the temporal and spatial scales of research better match those of management, but a fundamentally different, "postnormal" conception of science itself may also be required (Sayre et al. 2012).

Meanwhile, another alternative to conventional range science has complicated the situation still further. It also came from overseas, concurrently with the nonequilibrium model, and it, too, started from one of range science's blind spots. In the 1960s, Allan Savory, a wildlife biologist from British colonial Rhodesia (now Zimbabwe), derived livestock grazing strategies and management principles from observations of the grazing patterns of wild ungulates in the presence of predators in Africa (Savory and Parsons 1980). After leaving Zimbabwe in 1979, Savory moved to New Mexico and began promoting his ideas internationally under the names "short duration grazing," "the Savory Grazing Method," "holistic resource management," and later simply "holistic management." His model involves bunching livestock together into herds and moving them frequently, producing short periods of intensive impact followed by longer intervals of rest. (Clements himself made much the same observation with regard to bison on the Great Plains, and Savory claims that the "herd effect" accelerates succession.) In areas without many predators or herders, such as the United States, this often entails fencing rangelands into many, small pastures.

Although his ideas occasionally reached publication in the range science literature (Goodloe 1969; Savory 1978, 1983; Savory and Parsons 1980), Savory relied on private short courses and consultancies rather than conventional scientific or institutional channels, and he eventually reached large audiences and attained near-celebrity status based on the success stories of ranchers who embraced his ideas. But many range scientists dismiss the success stories as anecdotes, because scores of controlled experiments have concluded that Savory's ecological claims are demonstrably false (Briske et al. 2008). Especially in the United States, range scientists are alarmed that government policies for rangelands might be imposed on the basis of a scientifically flawed theory, both for the precedent it might set for other realms and, one might suspect, for the threat it poses to their own institutional authority. They are right to be dubious, but they might start by acknowledging that the problem is hardly new: the imposition of a scientifically flawed theory through government rangeland policies is the history of their own discipline, in which their predecessors were complicit, if not chiefly responsible. Holistic management's popularity raises the possibility that a scientifically flawed theory—*if willingly embraced rather than imposed*—may in some cases induce improvements in range management for reasons that cannot be—or at least have not yet been—reducible to controlled experimental testing (Briske et al. 2011).

CONCLUSION

Capital, Climate, and Community-Based Conservation

In stark contrast to forty years ago, today the theoretical basis of rangeland ecology is a topic of active debate and research. Compared to the mid-twentieth century, when US range scientists constituted a rather insular professional and scientific field, the participants in these debates are a diverse group, including modelers, theorists, and field scientists from around the world and across numerous subfields of ecology. The *Journal of Range Management* underwent a name change—to *Rangeland Ecology and Management*—in 2005 to reflect this broader, less livestock-focused identity. The importance of nonequilibrium ideas is not in dispute, but the details of those ideas, how they relate to older, equilibrium approaches, and how they should inform rangeland science, policy, and management are more or less open questions (Illius and O'Connor 1999; Vetter 2005). And unlike a century ago, when Frederick Coville casually equated rangelands with pastures, the question of scale is now explicit and central.

Conceptually, both equilibrium and nonequilibrium require a temporal scale of reference—a period of stability, or change relative to such a period—as well as a spatial unit or boundary of whatever it is that is taken to be in equilibrium (or not). Insofar as any actual ecosystem has components that change over different spatial and temporal scales, equilibrium and nonequilibrium are not mutually exclusive categories. The development of state-and-transition models still entails choosing some reference site or baseline to represent potential conditions on a given type of range, and changes within a state—those that do not cross a threshold of relatively permanent change—may still resemble succession as described by Clements. "Equilibrium and non-equilibrium ecosystems are not distinguished on the basis of unique processes or functions, but rather by the evaluation of system dynamics at various temporal and spatial scales" (Briske, Fuhlendorf, and Smeins 2003, 601).

In short, there is wide agreement that in both theory and practice, rangeland ecology must be conducted at and across multiple scales. The history presented in the preceding chapters shows that this has long been the case, but that a variety of factors obscured it from view until fairly recently. Some of these factors were "scientific" in nature, including the dominance of Clementsian theory (and the University of Nebraska) in the formative years of ecology and the chronic insufficiency of data for many rangelands. But the major obstacles—the sources of the blind spots in which the scale problem lurked—were institutional, political, and economic. The fact that an alternative theory has emerged does not mean that these factors have somehow gone away and ceded the field to "purely" scientific considerations. The broader social forces that condition the science of rangelands have changed, but they are still there, even if rangeland ecologists do not often discuss them in print.

Like any science, range science needed a theoretical basis in order to become a science at all: without a conceptual framework for organizing knowledge systematically, information gained in one time and place could not be related to other times and places. Clements (1916) himself proudly proclaimed successional theory to be "universal," a discovery of the "natural laws" of vegetation development and change, and it was instrumental as such to the emerging field of ecology. But range science was compelled to adopt a theory for extrascientific reasons, and it did so prematurely. From the outset, government administrators expected range scientists to produce unequivocal, predictive knowledge about rangelands in the belief that such knowledge would provide solutions to ecological and economic crises as well as impartial, apolitical "facts" with which to resolve disputes. But the methods, epistemologies, and objectives of science differed from those of politics and economics; knowledge, public support, and profit rarely aligned perfectly. The scientists were not dominated in any simple way—after all, without at least the appearance of some independence, scientific knowledge could not serve the purpose of providing authority (for political needs) nor insight (for economic ones)—but they were constrained in the questions they could ask and the resources available to answer them.

If having a theoretical basis was necessary to qualify as a science, being a science was necessary to meet the needs of government agencies as they contended with the demands of ranchers, capitalists, and the broader public. Within these complex constraints, range scientists *produced* rangelands in a twofold sense. Directly, they produced ideas about how rangelands worked and should be managed. Indirectly, these ideas informed laws, policies, programs, and management guidelines that altered rangelands across the western United States and subsequently elsewhere in the world, although the rangelands themselves often confounded or exceeded the data, methods, and interpretations available through scientific inquiry. One abiding feature—common to Merriam's calculations of prairie dog damages, Coville's predictions about predator-free sheep production, Jardine's forage acre, and decades of carrying capacity estimates—was the postulation of an imaginary ideal based on a model of how nature and/or the market were supposed to work. Empirical reality inevitably deviated from the ideal, but these disparities—especially if they could be quantified—lent the models a social potency out of proportion to the scientific observations and knowledge that ostensibly supported them. This form of reasoning is not unique to range science and did not originate within it; indeed, it is almost diagnostically modern and is arguably what has made science so central to projects of the state. Mistaking the model of reality for reality itself, it disguised normative abstractions as positive facts and then set about to make reality conform, whether by dictating management to ranchers and pastoralists, applying the brute force of machines and chemicals, or bureaucratic sleight of hand.

What confounded the models was less the West's celebrated aridity than its variability. Average annual precipitation is indeed important for understanding ecosystems, because at evolutionary and geological scales it makes some soils and organisms possible and others impossible. But at shorter and more practically relevant time scales, averages can hide as much as they reveal if the *actual* precipitation in a given year deviates widely from the long-term average. Scientists capture this by measuring the coefficient of variation: how much, on average, a given year's total differs from the long-term average. The lower the average rainfall, the higher the coefficient of variation is likely to be: in a site that normally gets forty inches, a four-inch deficit is only 10 percent, whereas in an area accustomed to only twelve inches, the same shortfall represents a 33 percent variation. Ecologists now believe that rangelands with a coefficient of variation of about 30 percent or greater have ecosystem dynamics that are qualitatively different from less variable ecosystems and that successional theory works only in the less variable settings (Illius and O'Connor 1999). By this rule of thumb, range science was right to embrace Clementsian theory for the Great Plains, where it originated, and perhaps for the northwestern states of Idaho, Washington, and Oregon, where Sampson did his pioneering research. But in large parts of the Southwest, stretching

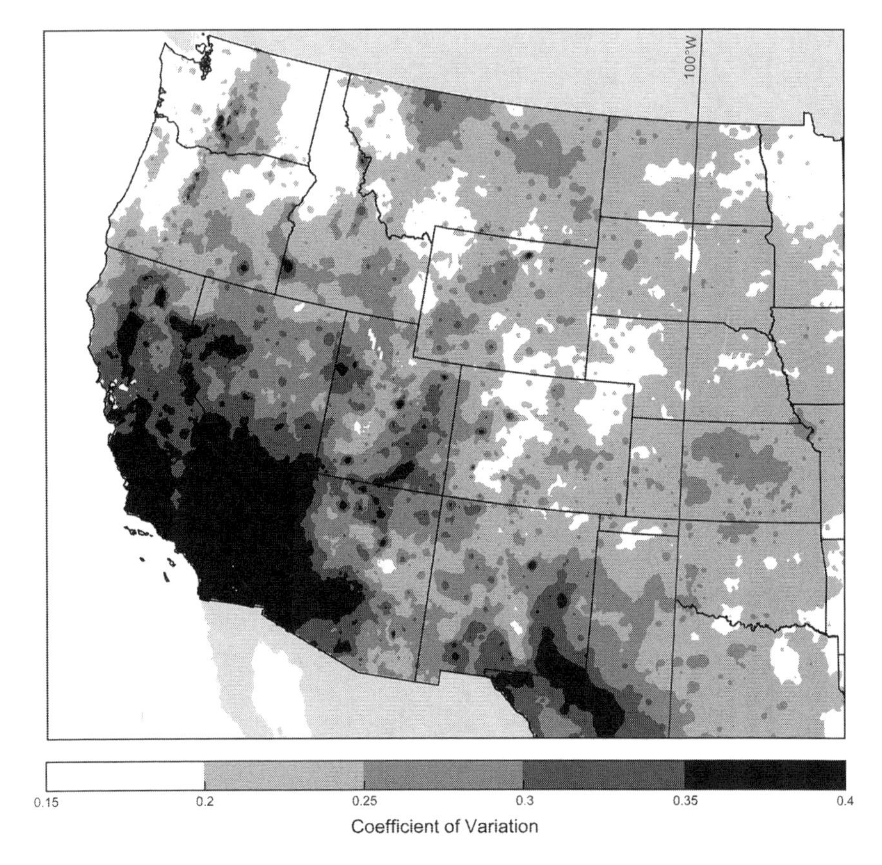

Map C.1. Coefficient of variation of annual precipitation in the western United States, 1950–2000. By N. L. Miller and S. Kim, using Matlab analysis code by N. L. Miller and S. Kim, based on data from Livneh et al. (2013).

from western Texas to Utah and California, greater variability made successional theory inappropriate and misleading (map C.1).

In retrospect, then, the source of the main problems for range science was the dominance of a national agency in its formative period between 1910 and about 1950. The Forest Service viewed uniform policies and management guidelines for all its lands as both bureaucratically and politically necessary. Two scale problems conspired: one of the temporal scale at which to evaluate rangeland ecosystems, the other of the spatial scale for applying the resulting knowledge. These problems were compounded by the fact that range science was subordinate within the agency to forestry—and therefore also to fire suppression—and bureaucratically separated from other agencies working in rangelands.

If scale was at the core of range science's missteps, however, capital and climate were the broader realities that made the missteps so fateful. Each imposed a temporal scale incompatible with the other. Mediated by debt and fluctuating market prices, capital demanded regular annual returns from ranchers and rangelands. But highly variable climate meant unpredictable booms and busts in forage and livestock production at the scale of years to decades. Government agencies turned to science for knowledge and techniques to reduce these risks, but experiments conducted at plot or pasture scales could not safely be extrapolated in any simple linear fashion to entire regions or landscapes, nor could results obtained over periods of a few years safely be assumed to hold all the time. Even recognizing this problem was difficult until data had been collected over the relevant spatial and temporal scales; methods of modeling ultimately enabled scientists to think at large extents without getting mired in the exponential variation of data collected at the grain of quadrats and transects. But state-and-transition models are still (only) models, and their predictive power remains to be demonstrated.

These are not merely academic or historical issues; they have had and continue to have real consequences. Markets and debt still impose contradictory temporal demands on ranchers, and the science of rangelands is still fraught with uncertainties. Take, for example, the seemingly simple question, "What is the current condition of the world's rangelands?" Many scientists have expended a great deal of effort to answer this question, using a variety of methods and data. But defining rangeland conditions, let alone measuring them, involves deep issues of ecological theory as well as extensive empirical data. Some scholars have concluded that 10 to 20 percent of rangelands are degraded; others place the number at 70 to 80 percent (Millennium Ecosystem Assessment 2005). One cannot safely split the difference, because what counts as "degradation" varies in kind as well as in number. Is it negative change from some past state? If so, when did the past state occur, and why should it serve as the baseline? What—and who—defines the change? In what sense is it negative? Is it reversible, and on what time scale? What—or who—caused it to happen? Scientists have not settled these questions among themselves, and meanwhile, their assessments of rangeland conditions, when deployed through governments and international programs, can still have dramatic and sometimes disastrous consequences for communities and landscapes. There are still places today, such as China, where the ranching model of fences and carrying capacities is being coercively implemented (Li and Huntsinger 2011).

The science of rangelands now faces radically different scientific, political, and economic circumstances than it did when it began. What appears to be a more robust theoretical framework has emerged, and land management and development agencies have begun to embrace it in some places, including the United States. Information at the scales of actual range management—real-time, high-resolution, over entire ranches and range landscapes—may soon be economically and technically available through remote sensing, unmanned aerial vehicles, and spatial analysis. Digital information technology and modeling make it feasible to manage and interpret the resulting data and to share it with anyone connected to the Internet. In these ways the scale of the problem is diminishing. But political and economic priorities have shifted. The western range—understood as the system of leases that links private ranch lands to fenced, publicly owned grazing allotments—was built on one foundational premise, expressly stated in the first page of the report of the second Public Lands Commission (Richards et al. 1905, 7, emphasis added): that rangelands "are, *and probably always must be*, of chief value for grazing." Secure, exclusive land tenure would therefore give ranchers incentive to conserve public and private rangelands alike, because their livelihoods would always depend on the forage. Today the premise no longer holds: the highest economic use of most western rangelands is for residential development (on private lands), tourism and recreation (on public lands), and conservation or energy development (on both). None of these uses is entirely independent of the climate, but all of them shift the circuit of accumulation to processes that are indifferent to interannual variability, decoupling economic activity from the ecological pulse of rangelands. And although ranchers are still powerful to a degree that is disproportionate to their numbers, they no longer have the political clout they once did, having been swamped by growing urban populations and widely vilified by environmentalists. The number of livestock on the western range has declined to the point that grazing generally cannot act as the major driver of vegetation change, but neither can it be the primary focus of capital accumulation and government policy.

Meanwhile, the reality and prospects of climate change pose radically new challenges now and for decades to come. The impacts of rising atmospheric greenhouse gas concentrations are already evident in the Southwest in severe droughts, declining snow packs and water levels in reservoirs, and extensive tree mortality and wildfires in forests throughout the region (MacDonald 2010; Seager and Vecchi 2010). Climate models suggest that variability will increase beyond its already high levels and that rising temperatures may make range livestock production effectively unfeasible by midcentury because, even if precipitation does not decline, evapo-transpiration will increase to the point that average soil moisture levels will fall within what was previously deemed severe drought (Seager et al. 2007; Chambers et al. 2015). In a cruel and supreme irony, there is also now evidence that the Sahelian drought that catapulted desertification to prominence in the 1970s may have been caused by sulfate aerosol and carbon dioxide emissions from the industrialized world (Ackerley et al. 2011).

What is at stake in the United States is not only the 300-odd million acres of publicly owned rangelands. It is well established by ecological research—and unsurprising given the history of land disposal—that the West's public lands are on balance drier, steeper, and less ecologically productive than its private lands, and that the public lands alone are insufficient to conserve biodiversity (Scott et al. 2001). The private lands are key, and they are rapidly succumbing to other land uses: in the intermountain West in the 1990s, roughly 1.6 million acres of rangelands ceased to be grazed every year, nearly half of it converted to urban uses (Sullins et al. 2002). As of 1998, some 21,000 individual ranch operators held leases to roughly 30,000 Forest Service and Bureau of Land Management grazing allotments; the private lands owned by those ranches totaled approximately 107 million acres. Although half of the ranches could be described as "hobby ranches," meaning their owners received most of their income from off-ranch sources, the other half depended for their livelihoods on their herds and, by extension, on continued access to public lands for forage (Gentner and Tanaka 2002). For all its flaws, the western range nonetheless succeeded in tying public and private land uses together in a way that dampens—imperfectly but importantly—what are now the greatest threats to rangelands in the United States and elsewhere: subdivision, fragmentation, and land-use change (Hansen et al. 2005; Galvin et al. 2008; Sayre et al. 2013).

Some critics of public lands grazing still cling to the Clementsian faith that removing livestock will cause all rangelands to restore themselves (Donahue 1999; Wuerthner and Matteson 2002). Overcoming this intuitive but erroneous idea is arguably the most important achievement that rangeland science has thus far produced. But while nonequilibrium ecology and better data may be necessary, they are not sufficient for successful range management. Technology can narrow the gap between plots and landscapes, range science and administration, small scales and large, but the problem of scale remains both scientifically and practically daunting. Threshold changes involving nonlinear interactions across multiple scales are still difficult to model or predict. Moreover, actual range management still takes place at multiple, intermediate scales, and it involves people, including but not limited to ranchers, whose interests and abilities may vary as widely as their landscapes.

What lessons can the history of range science offer for the present and future? Many of the blind spots that developed in the early decades of the discipline have since been revealed, but are there others yet to be exposed and explored? Part of the justification for range science circa 1900 was that US ranchers, as newcomers to the West, were ignorant of its landscapes and in need of scientific expertise to educate them (Chapline 1937). Today, many ranchers have generations of experience and knowledge on their particular pieces of rangeland, and scientists are coming to recognize that experienced managers have as much to teach as to learn (Knapp and Fernández-Giménez 2008, 2009a, 2009b). The same could, of course, be said about African and Asian pastoralists, who have developed their management practices over thousands of years. But such recognition should not be limited to "traditional" or "local" ecological knowledge, viewed as distinct and different from "scientific" knowledge. Recall the experiment conducted by H. L. Bentley outside Abilene, Texas, in the late 1890s, described in chapter 5. The goal was to improve the carrying capacity of the land being studied, which for obvious reasons required some method of measurement at the outset and after each year of treatments. There also needed to be a way to decide how many animals to place in each pasture. Rather than measuring plants or setting fixed stocking rates, Bentley chose a completely different method:

> Three well-known stockmen of central Texas were invited to make a full and painstaking inspection of practically every part of the section. They were C. W. Middleton, J. W. Parramore, and W. J. Bryan, all of Taylor County and all old settlers in that part of the State, each a large owner of cattle, and, therefore, specially interested in the results to be secured. That each one of them could accurately estimate the capacity of a range to sustain stock no one in the Southwest, where they are extensively known, could for a moment doubt. It was believed, therefore, that an expression of opinion by them on the subject would be accepted as definitely determining the capacity of the particular section under consideration. (Bentley 1902, 15)

Each subsequent year, the committee of local ranchers examined the pastures again and recommended a stocking rate for the coming year. Bentley effectively put them in charge of the experiment's main variable, allowing them to change it in response to fluctuating conditions. E. O. Wooton echoed this approach when he envisioned the Jornada as a place where Charles Turney would "run the ranch as he would do if it were his own land," and the scientists would simply study the results. It was a much deeper kind of cooperation than using a rancher's herd in an experiment designed by scientists, as happened on the Jornada after Turney departed in 1925 and still happens today on the Santa Rita Experimental Range. The reliance on nonscientists as experts and the flexibility to adjust the experimental design rather than strictly following "the definite plans for treating the station pastures as originally laid down" (Bentley 1902, 22) were virtually unheard of in range science for the subsequent half century or more, because both threatened the control deemed necessary for quantitative, "scientific" analysis and hypothesis testing. It is precisely this flexibility, though, that variable rangelands apparently demand. It is also why Allan Savory insists—perhaps correctly—that his method of grazing management cannot be tested experimentally (Savory and Parsons 1980; Briske et al. 2011). Even though stocking rates were confounded with other variables in Bentley's study, something like this level of collaboration might have set range science on a different course from the beginning.

The land tenure arrangements of the western range—produced by fences and leases and abetted by predator control—represent another potential blind spot, given that both history and theory suggest the importance of approaching rangelands at scales larger than an individual ranch or grazing allotment. Research from pastoralist settings confirms this insight: "Dynamic ecological systems require tenure systems to be flexible, to allow reciprocal access in bad years, to allow mobility on varying scales according to the conditions of the year, to allow for highly variable scales of exploitation and to include detailed provision for access to and management of key resources" (Swift 1994, 160).

Research at such scales was rare in the history of US range science, even on the large experimental ranges (Sayre et al. 2012), but it was neither unknown nor impossible. Beginning in 1926, for example, an experiment took place on 108,000 acres of degraded rangelands in southeastern Montana known as the Mizpah-Pumpkin Creek Grazing District. Land ownership there was a complex mosaic of parcels belonging to the Northern Pacific Railway, numerous private ranchers, the state of Montana, and the Department of the Interior, the last being unclaimed public domain. Some fifteen ranchers conceived and led the project as a pilot, formally authorized by Congress in 1928 to demonstrate the viability of grazing districts on the public domain. The entire area was grazed in common, with a hired cowboy minding the herd, mending the perimeter fence, and poisoning prairie dogs. Decisions related to labor, water development, bull selection and acquisition, and other inputs were all made collectively. A county agricultural extension agent facilitated the organization, and a Forest Service grazing specialist was consulted at the outset, but otherwise the involvement of trained experts was minimal. "Except for the slight supervision which is given by the representative of the Interior Department, management of the Mizpah-Pumpkin Creek grazing district is entirely by the ranchers who are members. . . . Much of the success of this grazing area is due to [the] high type of leadership and membership in the association, their willingness to submerge the cattleman's traditional individualism in favor of group benefits, and to make use of various existing cooperating agencies" (Tootell 1932, 6). Although the area of the Mizpah-Pumpkin Creek Grazing District was not very large by Western standards, the collective use of the range allowed management to operate across ownership boundaries, at larger scales than the ranchers could have done individually. Range conditions improved markedly compared to the surrounding public domain, and the success helped persuade livestock associations to support the Taylor Grazing Act of 1934, pleased that "the Mizpah-Pumpkin Creek Grazing Association had been allowed to administer their own affairs with little interference from the Interior Department" (Muhn 1987, 80).

Both of these experiments are now nearly forgotten, but they point to what might have happened if the practices of range science had been institutionalized differently. The Taylor Grazing Act created grazing districts throughout the Department of the Interior's rangelands, and most were significantly larger than Mizpah-Pumpkin Creek. But they merely coordinated administration among local groups of ranches without pooling their management, and ranchers' authority to make management decisions was a chronic point of contention. The first head of the Grazing Division, Farrington Carpenter, envisioned empowering ranchers to make many decisions through district advisory committees, but for critics this smacked of co-optation—the foxes guarding the henhouse, so to speak—and it ran afoul of Interior Secretary Harold Ickes, who forced Carpenter to resign in 1938 (Merrill 2002). In the 1950s, the Bureau of Land Management feuded with the Mizpah-Pumpkin Creek Grazing Association, mainly over the issue of who would determine stocking rates (Muhn 1987, 97), and the association's lease was canceled in 1962 to harmonize with the national Taylor Act template. Among range scientists, the creation of experimental ranges and other dedicated research sites discouraged collaboration with landowners by providing places to do their experiments without people, who—like the herders for Jardine during the Coyote-Proof Pasture Experiment—might introduce variables that could not easily be measured or controlled.

The spatial and intellectual abstraction of rangelands from people was part of becoming "scientific," but in retrospect it created a blind spot that has only recently been acknowledged. As with Clementsian theory, the core problem was variability: Highly variable systems require long time scales for experiments to distinguish between statistical noise and valid signals, but "the longer that experiments run without the ability to respond to changing conditions, the more likely they are to deviate from what would occur on working ranches, where managers respond flexibly, in real time, to the inherent spatial and temporal variability of rangeland ecosystems" (Briske et al. 2011, 328). In recent years, range scientists have begun to recognize that "the human processes are as important as the ecological processes" in determining management effects and that managers' "mental models" of how rangeland systems work "are a crucial slow variable" that warrants as much study as the rangelands themselves (Lynam and Stafford Smith 2004, 71, 76). "Rangeland landscapes are extremely heterogeneous; general principles derived from scientific experimentation cannot be easily or generally applied without adjusting to the distinct social and ecological characteristics of a location" (Sayre et al. 2012, 545). Ironically, skilled, experienced herders may offer the best way to achieve the highly site-specific goals that society now has for rangelands in the United States.

Public land agencies are also beginning to recognize the importance of social factors. The idea of adaptive management—which shares roots with the nonequilibrium model (Holling 1978)—has gained adherents among scientists and agencies alike. It calls for working at large scales and treating natural resource management as an ongoing experiment, involving all concerned parties in an iterative process of "learning by doing" (Walters and Holling 1990). On this understanding, "command and control" management that imposes static prescriptions tends to degrade the capacity of both social and ecological systems to cope effectively with subsequent disturbances and variability (Holling and Meffe 1996). Whether the adaptive approach can be reconciled with bureaucratic norms and habits remains to be seen, but the Forest Service has revamped its rangeland monitoring procedures to place much greater emphasis on ensuring that ranchers participate directly in collecting data. The agency even allows them to do it themselves, because this simple step appears to improve management through its effects on the ranchers' mental models—indeed, this may be effectively more valuable than what the ecological data appear to indicate about a grazing allotment (Sayre, Biber, and Marchesi 2013).

Perhaps most encouraging, in this light, is the emergence in places throughout the West of community-based conservation efforts, initiated at the scale of watersheds and landscapes by groups of landowners in cooperation with agencies, environmentalists, range scientists, and experts in related social and biophysical sciences (Charnley, Sheridan, and Nabhan 2014). The goals and motivations of these groups vary depending on local circumstances, but most share concerns about declining rural economic prospects, the specter of land-use change and fragmentation, and the inflexibility and complexity of environmental regulations and policies within and across agencies and jurisdictions. They seek to address threats and problems on both public and private lands, not only for economic benefit but for environmental, aesthetic, and cultural reasons. Community-based conservation efforts led by ranchers have succeeded in expanding the scale of grazing management across Forest Service and Bureau of Land Management allotments in the West Elk Mountains of western Colorado (Bradford 1998), restoring fire to federal, state, and private rangelands in the Malpai Borderlands of Arizona and New Mexico (Sayre 2005c), and developing methods for monitoring both the ecological and socioeconomic conditions of rangelands in northern Arizona (Munoz-Erickson, Aguilar-Gonzalez, and Sisk 2007). Many include or cooperate with nonprofit land trusts to secure conservation easements that prevent subdivision and development of private lands (Rissman and Sayre 2012), and a handful work closely with scientists to conduct research aimed at resolving acute management or regulatory problems. They face countless obstacles, and there is no guarantee that they will succeed in their goals. But they at least begin to construct ways of governing rangelands at the large spatial and long temporal scales that are—as best as we understand—the right ones. And they do so not by invoking the rigid power of the state, the unilateral authority of science, or the faceless abstractions of the market but by acknowledging that the management of land, like the production of knowledge, is a social process. Its outcomes do not depend simply on the "correct" application of "the best science" but on the qualities of the process itself: how people participate and interact, their experiences and aspirations, and their humility—or lack thereof—in the face of lands that are not only unpredictable but still far more complex than we understand.

NOTES

INTRODUCTION

1. The extent of rangelands depends on one's definition and data sources; 40 percent of Earth's ice-free land—roughly 56 million square kilometers—is a reasonable round figure (Asner et al. 2004; Reid, Galvin, and Kruska 2008). Estimates are complicated in part by the fact that rangelands overlap but are not identical with categories such as arid lands, drylands, wild lands, and grazing lands.
2. Estimates of the proportion of rangelands that are grazed vary widely. Asner et al. (2004) calculate that "managed grazing" occupies 33 million square kilometers of land, and that in all biomes, grazing occurs on no more than 56 percent of the total area. Reid, Galvin, and Kruska (2008) put the number at nearly twice that amount: 61.2 million square kilometers, of which 91 percent is "extensive grazing lands," roughly equivalent to rangelands. Holechek, Pieper, and Herbel (2004) estimate that 70 percent of Earth's land (excluding Antarctica) is "potentially grazable."
3. "Rangelands occur in areas of relatively low rainfall or where winters are long and cold" (Grice and Hodgkinson 2002, 2).
4. Fragmentation is a threat to extensive grazing lands—a category encompassed by but not identical to rangelands (Galvin et al. 2008). The Millennium Ecosystem Assessment (2005, 13) identifies drylands—a category that mostly overlaps with rangelands—as particularly at risk due to high rates of population growth, low biological and economic productivity, limited water supplies, and highly variable climate.
5. The 1999 chapter recapitulates and extends a chapter in Rowley's earlier book (1985). Worster (1985) devoted a chapter of his history of ecological ideas to Frederic Clements, but he did not discuss range science. A few more local histories do exist: Alagona (2008) explores the history of scientific ideas about California's oak savanna rangelands, and Godfrey (2013), Shapiro (2014), and Alexander (1987a, 1987b) look at specific Forest Service range research stations.
6. See Campbell (1948); Chapline (1937, 1948, 1951, 1980); Holechek (1981); Joyce (1993); Owensby (2000); Talbot and Cronemiller (1961); Young and Allen (1997); Olberding, Mitchell, and Moore (2005); Prevedel and Johnson (2005); and Prevedel, McArthur, and Johnson (2005). Young (2000) and Young and Clements (2001) published short articles on range science up to 1930, as well as a monograph on cheatgrass that contains some historical chapters (Young and Clements 2009), but

they chose to exclude the Southwest, where the earliest range research stations were located.

CHAPTER ONE

1. Insects—including parasites such as ticks and disease vectors such as screwworm flies—were also major targets of government extermination efforts, carried out by still other USDA agencies such as the Bureau of Economic Entomology. These efforts are beyond my scope here, and they are less relevant in that they had minimal impacts on the ecology of rangelands (as far as anyone knows).
2. Smith (1898) describes what is probably the first such experiment, at least the first by government researchers. See chapter 5.
3. Jared Smith (1899, 14–15) made many of the same arguments two years earlier.
4. The methods behind this estimate appear to have been extremely crude. The portion attributed to house rats—$200 million—probably came from Lantz (1918, 236), who effectively pulled it out of thin air: "Assuming that there are in the United States only as many rats as people, and that each rat in a year destroys property valued at $2, the total yearly damage is about $200,000,000."
5. The growing dominance of economic over scientific priorities subsequently drove Merriam to resign his position as chief of the survey in 1910 and decamp to an endowed research position with the Smithsonian Institution (Osgood 1944).
6. In 1929, the Division of Predator and Rodent Control (PARC) was established within the National Biological Survey; ten years later, PARC was transferred to the Interior Department as part of the recently created US Fish and Wildlife Service, where it remained (though renamed Animal Damage Control) until 1986, when it was returned to USDA under the name of the Animal and Plant Health Inspection Service.
7. According to internal reports in the National Archives, CCC camps run by the Interior Department's Division of Grazing had completed rodent control on 6,645,889 acres of land in the West as of October 31, 1938. This total does not include similar efforts on national forest lands, nor does it include further work conducted from 1938 to 1942, when the CCC was disbanded.
8. The Imnaha National Forest was established on March 1, 1907, by the combination of the Wallowa and Chenismus National Forests, which had been established by President Roosevelt in 1905. On July 1, 1908, the name was changed to Wallowa National Forest, and in 1954 it was combined with the Whitman National Forest to create the Wallowa-Whitman National Forest (Davis 1983, vol. 2, 743–88).
9. Merriam's memo was accompanied by specifications for the proposed fence prepared by Biological Survey employee Vernon Bailey, who was also Merriam's brother-in-law and longtime specimen collector (Jardine 1908). Bailey's designs for predator-proof fencing are detailed in Bailey (1907).
10. Quotations are from Coville's handwritten notes, found in the records of the Bureau of Plant Industry (Record Group 54) in the National Archives and Records Administration, College Park, Maryland.
11. Coville may also have been influenced by reported results of fencing out predators to protect sheep in Australia and South Africa. In a 1905 Bulletin of the Biological Survey, David Lantz (1905, 23) gave a nearly identical list of advantages, beginning with "decreased cost of herding," from a paper read by the president of a farmer's association in Cape Colony, South Africa.
12. Although the Wallowa experiment did not involve lambing—that is, the handling of ewes and lambs during the birthing (or yeaning) period—Jardine took a keen

interest in the question of whether the pasturage system would also work for this purpose. At first he relied on reports from private sheep owners who had implemented variations along similar lines on their own (Jardine 1909, 1910, 1911); later he oversaw an experiment in the Cochetopa National Forest in Colorado (Jardine 1911). I have not treated these in detail here because the conclusions he reached were so similar to those reached in the Wallowa experiment. On the critical issue of economic returns, for example, the conclusion was identical: at 8 percent interest, a coyote-proof lambing pasture would pay for itself and produce a dividend after six years (Jardine 1911, 32).

13. According to USDA (2010, 2011) data, cattle losses from animal predators in 2010 totaled 219,900 head, more than half attributed to coyotes. Although valued at nearly $100 million, this represented less than one-quarter of 1 percent of the national herd. Eighty-two percent of losses were calves. Predation by dogs exceeded that by wolves by nearly 270 percent (21,800 to 8,100). Predation of sheep totaled 247,200 head; of goats, 180,000 head. In the eleven western states, cattle and calf losses totaled 55,000 head, and sheep and lamb losses totaled 119,700.
14. In the thirteen national forests in the southwest region of the Forest Service, for example, 4,153 miles of fence were constructed between 1925 and 1933, at a total cost of $851,000 (more than $12 million in 2015 dollars); the agency paid 49 percent of this cost, while ranchers paid the rest.
15. Cowden (1959) reported the following range improvements funded by the Agricultural Adjustment Act in Arizona alone: "thousands of miles of fence" constructed, 900 springs developed, 8,000 earthen reservoirs built, 1,300 wells drilled, 600 miles of pipelines installed, and spreader dams (a kind of erosion-control structure) containing 13 million cubic yards of earth. Federal lands were not eligible for these funds, so all these projects were in addition to activities by the Forest Service and the Division of Grazing/Bureau of Land Management.
16. This figure includes all CCC crews, not only those working on national forests, federal lands, or rangelands. Rangelands were a major site of CCC work, however. "The range improvement program [of the Grazing Division of the General Land Office] was primarily dependent on the use of the Civilian Conservation Corps. In the first year of its operations, the Division of Grazing was assigned sixty CCC camps, with about 12,000 men, to carry out range improvements on the public domain. Use of the CCC was continued until 1943, when the corps was discontinued. At one time as many as one hundred camps were assigned to the Division of Grazing. Very likely most of the range improvements constructed since the inception of the [Taylor Grazing] act were accomplished by the CCC. The availability of CCC labor was a most fortunate circumstance for the infant Division of Grazing" (Foss 1960b, 82).
17. For example, replacing the top and bottom strands of barbed-wire fences with smooth wire is a recognized "wildlife-friendly" practice supported by government conservation programs, but predators are not really the intended beneficiaries. Meanwhile, the protection or reintroduction of predators has gained traction—and provoked controversy—in recent decades, but fence removal has not been proposed as a way of advancing the cause.

CHAPTER TWO

1. A forest ranger on the Coronado National Forest included these lines in a memo about the Cottonwood Grazing Allotment in 1932: "There is very little to be said about this range except that it is in excellent condition[,] is under good management

and well cared for by the permittees. We have a fire hazard there now." Similarly, in a pair of memos from 1928—the second labeled "confidential"—Inspector of Grazing D. A. Shoemaker criticized a district forester for poor grazing management on many allotments, which had produced dangerously high fuel loads. The problem was not too few livestock, Shoemaker wrote, but rather their poor distribution. "We cannot be committed to a policy or a practice that will damage ranges, watersheds or timber by overgrazing as a means of reducing fire hazards."

2. At the District 3 Grazing Studies Conference in December 1921, Enoch Nelson from the Jornada reported on tobosa grass range that had been burned in 1919 and 1921: "The apparent results secured was at least 30% reduction in density, reduced height, growth, and consequently volume." Despite the reduction in density, however, cattle utilized the post-fire forage more. "Further study on the permanent effect of burning on density is contemplated" (US Forest Service 1922, 94).
3. Sampson's obituary in the *Journal of Range Management* (Parker et al. 1967, 346) asserts that he did postgraduate work at Johns Hopkins in 1914–1915, but the Johns Hopkins Registrar's Office reports no record of his enrollment there.
4. The closest anyone has come is Tobey's (1981) sociological study of the founding of American Plant Ecology, which treats Clements in some detail but does not attempt a biography of him. Clements's wife, Edith, published an account of their life together after his death (Clements 1960), but it is really a memoir rather than a biography. Neither book contains any mention of Sampson.
5. "Clements had worked out his theory in the western United States, in landscapes that he viewed (along with most other Euro-Americans) as pristine, wild, or untouched by humans. Tansley, by contrast, took his bearings from the English countryside" (Sayre 2012, 59).
6. In his preface to David Griffiths's 1904 Bureau of Plant Industry Bulletin, "Range Investigations in Arizona," agrostologist W. J. Spillman wrote: "A knowledge of the carrying capacity of the ranges is of utmost importance, for it must form the basis of any intelligent legislation relating to the range question. This knowledge determines the rental and sale value of range lands and should also determine the size of the minimum lease or homestead for range purposes in case laws are passed providing for such disposal of the public ranges" (Griffiths 1904, 5).
7. Pyne (1997, 287) is referring specifically to the Southwest in this passage, but the point could be extended to many other parts of the West as well.
8. In 1951, for example, when Forest Service researchers were developing mechanical means of removing sagebrush (see chapter 5), Assistant USFS Chief E. W. Loveridge concluded a memo to the chief of the Division of Range Research by writing: "I presume it is felt to be too dangerous to suggest the practice advocated, I believe, in the Intermountain Region of controlled burning of sagebrush under certain conditions—and thereby saving the cost of mechanical removal of that type of competition."
9. Humphrey transferred from the Forest Service to the Soil Conservation Service sometime between 1937 and 1941.
10. Humphrey also collected data from burned and unburned areas two years after a small fire west of the Santa Rita, in the Sierrita Mountains. But the fire appears to have been natural or accidental, as evidenced by the absence of prefire data and by the vagueness of his description: "35 to 40 acres" in size, it burned "in April or May, 1933" (Humphrey 1949, 178).

CHAPTER THREE

1. Letter to Fred Ares, November 2, 1936, Jornada Experimental Range archives.
2. The contract with Turney also included language indicating another motivation on the part of the government: research was to be conducted on the "breeding of horses for military purposes."
3. Turney's exclusive access to graze on the Jornada Reserve also troubled Galloway, "because it appears to confirm monopolistic control of a large area of land by one individual for a long period. It would seem that the resulting public impression brought about by this action might be unfortunate in view of what Congress is attempting to do in the way of regulating grazing on the public lands."

CHAPTER FOUR

1. The idea of a reference site has been used recently in much the same fashion in connection with new methods of rangeland monitoring, although the term now often refers to a place that has never, or not recently, been grazed. The idea of using such sites to determine climax conditions was advanced by Clements (1934), who called them "relict sites."
2. The total area of Division of Grazing lands had been increased by Congress from 128 to 142 million acres in 1936 (Penny and Clawson 1953, 25).
3. Vesk and Westoby (2001) reviewed the published literature on 829 species that had been categorized as increasers or decreasers on Australian rangelands, and of those studied more than once, 41 percent were inconsistent—that is, they were *both* increasers and decreasers. They concluded that context, rather than genetics, determines many plant species' responses to grazing.

CHAPTER FIVE

1. Hayden was elected Arizona's first congressional representative in 1912, and he advocated for the Forest Service in one of his first speeches in the House. He became a senator in 1926, and he chaired the Appropriations Committee from 1955 to 1969, when he retired. His record as the longest serving member of Congress stood until 2009.
2. The origin of this estimate is unclear. In *The Western Range*, the area classified as extremely depleted (76 to 100 percent loss of forage production from "virgin" conditions) was estimated at 120 million acres (Secretary of Agriculture 1936, 4).
3. The others were weeping lovegrass (*Eragrostis curvula*) and Boer lovegrass (*E. chloromelas*); the latter has since been reclassified as a variety of the former (*E. curvula* var. *conferta*).
4. Notes from the committee's 1951 meeting indicate that whenever possible, the SCS tried to turn over seed production of highly sought varieties to the private sector, where agronomic production could take advantage of economies of scale: "Large-scale seed production by Government agencies on 'wild' lands is usually more expensive and less efficient than commercial production on farm lands. It is also felt that the Government should not compete with private growers in seed production."
5. Aerial seeding was attempted by the Bureau of Indian Affairs on 90,000 acres on four Indian reservations in Arizona in 1946–1948, using seeds of several native and (mostly) nonnative grasses encased in pellets of clay, fertilizer, and rodent repellent. The pellets did not penetrate the ground, however, and all the trials were deemed failures, attributed in part to the fact that "rats gathered up the pellets containing

the repellent and ate the seed with apparent relish" (Wagner 1949, 633). It was concluded that seedbed preparation was necessary prior to aerial seeding to soften the ground and enable penetration. Other aerial seeding experiments, on Bureau of Land Management and Forest Service lands in 1948–1949, showed some success with unpelleted seed, but follow-up measurements six years later found that all the gains had been lost (Bleak and Hull 1958).

6. When political opposition to the program grew in the United States in the late 1960s, Fred Tschirley was sent to Vietnam to assess its ecological effects. He reported that a single application on upland forests would have no "great or lasting effect," but that "a second application, especially if made within 3 or 4 months after the first, would have a wholly different effect" (Tschirley 1968, 52) by killing canopy trees, which would be replaced by bamboo. The bamboo, in turn, would interrupt succession back toward forest conditions. How long it would take for the forests to recover, he wrote, was unknown.

CHAPTER SIX

1. This is not to say that no one overseas had attempted to bring scientific methods to bear on range livestock production, but only that their work had not coalesced into a recognized or institutionalized field of science. Apart from work by anthropologists, the scholarly literature on range livestock production overseas at this time consisted mainly of reports by colonial agricultural administrators, for whom the topic was distinctly secondary to forestry, crop agriculture, and more intensive livestock production, including dairy. A few estimates of livestock carrying capacity had been made for pastoral areas in sub-Saharan Africa (Allan 1965, 291–94), and concerns about overstocking in Tanganyika Territory had been voiced by Hornby (1936), who advocated imposing rotational grazing schemes to prevent erosion and the loss of carrying capacity. New Zealand also had a network of research stations dedicated to converting its landscapes to introduced grasses for pasturing livestock (Brooking, Hodge, and Wood 2002).
2. The idea of calculating actual and potential human carrying capacities based on ecological data (e.g., soils and vegetation) was also developed by William A. Allan (1949) of the British colonial agricultural administration in Northern Rhodesia (present-day Zambia) for use in forcibly relocating native Africans (Sayre 2008). Vogt and Osborn do not appear to have been aware of Allan's work, however.
3. For example, "the science and art of planning and directing range use so as to obtain the maximum livestock production consistent with conservation of range resources" (Stoddart and Smith 1943, 2), or "the science and art of optimizing the returns from rangelands in those combinations most desired by and suitable to society through the manipulation of range ecosystems" (Stoddart, Smith, and Box 1975, 3).

CHAPTER SEVEN

1. Sandford wrote a blistering internal report for the World Bank (Sandford 1981) that was particularly influential. I am indebted to Roy Behnke for bringing this report to my attention and providing me with a copy of it.
2. The other three biome program areas were Coniferous Forest, Eastern Deciduous Forest, and Arctic and Alpine (Coleman 2010).
3. Another important but short-lived effort to use systems modeling to understand pastoralism was undertaken at Massachusetts Institute of Technology in 1973–1974 under a $1 million grant from USAID, part of a $30 million congressional appro-

priation to respond to the Sahelian disaster. Even before the one-year grant term had ended, administrators at USAID had apparently concluded that the effort would not meet their needs (Taylor 1992), and it does not appear to have had much effect on range science or pastoral development.

4. The claims and quotations in this paragraph are from an interview with Mark Westoby conducted by the author, January 26, 2015.
5. Author interview with Brian Walker, February 25, 2015.
6. E-mail communication from Mark Westoby, October 11, 2015.

BIBLIOGRAPHY

A Note on Archival Sources

Materials quoted without references in the text are from archival sources, of which five are most important: (1) the National Archives in College Park, Maryland; (2) the National Archives regional office in Denver, Colorado; (3) the archives of the Jornada Experimental Range in Las Cruces, New Mexico; (4) the archives of the Santa Rita Experimental Range in Tucson, Arizona; (5) the archives of the Coronado National Forest, also in Tucson. In the National Archives in both College Park and Denver, the sources used for this book are housed primarily in the records of the Forest Service (Record Group 95) and the records of the Bureau of Land Management and its predecessors, the Division of Grazing and the Grazing Division (Record Group 49). Materials related to the Division of Agrostology and the Coyote-Proof Pasture Experiment, including Frederick Coville's papers, are from the records of the Bureau of Plant Industry (Record Group 54).

Because the Jornada, Santa Rita, and Coronado National Forest archives are not cataloged, and the National Archives materials involve cumbersome cataloging systems, I have chosen to make these materials available by request rather than inserting identifying information for each document in the text.

Cited Sources

Abruzzi, William S. 1995. "The Social and Ecological Consequences of Early Cattle Ranching in the Little Colorado River Basin." *Human Ecology* 23:75–98.

Ackerley, Duncan, Ben B. B. Booth, Sylvia H. E. Knight, Eleanor J. Highwood, David J. Frame, Myles R. Allen, and David P. Rowell. 2011. "Sensitivity of Twentieth-Century Sahel Rainfall to Sulfate Aerosol and CO_2 Forcing." *Journal of Climate* 24:4999–5014.

Alagona, Peter S. 2008. "Homes on the Range: Cooperative Conservation and Environmental Change on California's Privately Owned Hardwood Rangelands." *Environmental History* 13:325–49.

Alexander, Thomas G. 1987a. "From Rule of Thumb to Scientific Range Management: The Case of the Intermountain Region of the Forest Service." *Western Historical Quarterly* 18:409–28.

———. 1987b. *The Rise of Multiple-Use Management in the Intermountain West: A History of Region 4 of the Forest Service*. Vol. 399. Washington, DC: USDA Forest Service.

Allan, William. 1949. *Studies in African Land Usage in Northern Rhodesia*. Rhodes-Livingstone Papers Number 15. Cape Town: Oxford University Press.

———. 1965. *The African Husbandman*. Edinburgh: Oliver & Boyd.

Alvord, Benjamin. 1883. "Winter Grazing in the Rocky Mountains." *Journal of the American Geographical Society of New York* 14:257–88.

American National Live Stock Association. 1938. *If and When It Rains: The Stockman's View of the Range Question*. Denver, CO: ANLSA.

Anderson, Darwin, Louis P. Hamilton, Hudson G. Reynolds, and Robert R. Humphrey. 1957. *Reseeding Desert Grassland Ranges in Southern Arizona*. Bulletin no. 249. Tucson: Arizona Experiment Station.

Archer, S. R., Kirk W. Davies, Timothy E. Fulbright, Kirk C. McDaniel, Bradford P. Wilcox, K. I. Predick, and D. D. Briske. 2011. "Brush Management as a Rangeland Conservation Strategy: A Critical Evaluation." In *Conservation Benefits of Rangeland Practices: Assessment, Recommendations, and Knowledge Gaps*, edited by David D. Briske, 105–70. Washington, DC: USDA Natural Resources Conservation Service.

Ares, Fred N. 1974. *The Jornada Experimental Range: An Epoch in the Era of Southwestern Range Management*. Edited by Robert S. Campbell. Denver: Society for Range Management.

Aronova, Elena, Karen S. Baker, and Naomi Oreskes. 2010. "Big Science and Big Data in Biology: From the International Geophysical Year through the International Biological Program to the Long Term Ecological Research (LTER) Network, 1957–Present." *Historical Studies in the Natural Sciences* 40:183–224.

Asner, Gregory P., Andrew J. Elmore, Lydia P. Olander, Roberta E. Martin, and A. Thomas Harris. 2004. "Grazing Systems, Ecosystem Responses, and Global Change." *Annual Review of Environmental Resources* 29:261–99.

Atherton, Lewis. 1961. *The Cattle Kings* Bloomington: Indiana University Press.

Bahre, Conrad J., and Marlyn L. Shelton. 1996. "Rangeland Destruction: Cattle and Drought in Southeastern Arizona at the Turn of the Century." *Journal of the Southwest* 38:1–22.

Bailey, Vernon. 1907. *Wolves in Relation to Stock, Game, and the National Forest Reserves*. USDA Forest Service Bulletin no. 72. Washington, DC: Government Printing Office.

Baker, Randall. 1984. "Protecting the Environment against the Poor: The Historical Roots of Soil Erosion Orthodoxy in the Third World." *The Ecologist* 14:53–60.

Barkema, Alan, Mark Drabenstott, and Nancy Novack. 2001. "The New U.S. Meat Industry." *Economic Review–Federal Reserve Bank of Kansas City* 86:33–56.

Bartlett, Ichabod S. 1918. *History of Wyoming*. Vol. 2. Chicago: S. J. Clarke.

Behnke, Roy H., Ian Scoones, and Carol Kerven, eds. 1993. *Range Ecology at Disequilibrium: New Models of Natural Variability and Pastoral Adaptation in African Savannas*. London: Overseas Development Institute.

Bell, W. B. 1921a. "Death to the Rodents." In *Yearbook of the Department of Agriculture 1920*, 421–38. Washington, DC: USDA.

———. 1921b. "Hunting Down Stock Killers." In *Yearbook of the Department of Agriculture 1920*, 289–300. Washington, DC: USDA.

Bennett, Hugh H., and William Ridgely Chapline. 1928. *Soil Erosion: A National Menace*. Circular no. 33. Washington, DC: USDA.

Bennett, John W. 1984. *Political Ecology and Development Projects Affecting Pastoralist Peoples in East Africa*. Research Paper, no. 80. Madison: Land Tenure Center, University of Wisconsin.

Bentley, H. L. 1898. *A Report upon the Grasses and Forage Plants of Central Texas*. USDA Division of Agrostology Bulletin 10. Washington, DC: Government Printing Office.

———. 1902. *Experiments in Range Improvement in Central Texas*. USDA Bureau of Plant Industry Bulletin 13. Washington, DC: Government Printing Office.

Bestelmeyer, Brandon T., Joel R. Brown, Kris M. Havstad, Robert Alexander, George Chavez, and Jeffrey E. Herrick. 2003. "Development and Use of State-and-Transition Models for Rangelands." *Journal of Range Management* 56:114–26.

Bleak, A. T., and A. C. Hull, Jr. 1958. "Seeding Pelleted and Unpelleted Seed on Four Range Types." *Journal of Range Management* 11:28–33.

Blydenstein, John, C. Roger Hungerford, Gerald I. Day, and R. R. Humphrey. 1957. "Effect of Domestic Livestock Exclusion on Vegetation in the Sonoran Desert." *Ecology* 38:522–26.

Bovey, Rodney W. 2001. *Woody Plants and Woody Plant Management: Ecology, Safety, and Environmental Impact*. New York: Marcel Dekker.

Box, Thadis W. 1992. "Rangelands, Desertification, and Clements' Ghost: A Viewpoint Paper." *Rangelands* 14:329–31.

Bradford, David. 1998. "Holistic Resource Management in the West Elks—Why It Works." *Rangelands* 20:6–9.

Brenner, Neil. 2000. "The Urban Question as a Scale Question: Reflections on Henri Lefebvre, Urban Theory and the Politics of Scale." *International Journal of Urban and Regional Research* 24:361–78.

Bridges, J. O. 1941. *Reseeding Trials on Arid Range Land*. Bulletin no. 278. State College: New Mexico College of Agriculture and Mechanic Arts.

Brisbin, James S. 1881. *The Beef Bonanza; Or, How to Get Rich on the Plains. Being a Description of Cattle-Growing, Sheep-Farming, Horse-Raising, and Dairying in the West*. Philadelphia: Lippincott.

Briske, David D., J. D. Derner, J. R. Brown, S. D. Fuhlendorf, W. R. Teague, K. M. Havstad, R. Li Gillen, Andrew J. Ash, and W. D. Willms. 2008. "Rotational Grazing on Rangelands: Reconciliation of Perception and Experimental Evidence." *Rangeland Ecology and Management* 61:3–17.

Briske, David D., Samuel D. Fuhlendorf, and Fred E. Smeins. 2003. "Vegetation Dynamics on Rangelands: A Critique of the Current Paradigms." *Journal of Applied Ecology* 40:601–14.

Briske, David D., Nathan F. Sayre, Lynn Huntsinger, Maria Fernandez-Gimenez, Bob Budd, and Justin D. Derner. 2011. "Origin, Persistence, and Resolution of the Rotational Grazing Debate: Integrating Human Dimensions into Rangeland Research." *Rangeland Ecology and Management* 64:325–34.

Brooking, Tom, Robin Hodge, and Vaughan Wood. 2002. "The Grasslands Revolution Reconsidered." In *Environmental Histories of New Zealand*, edited by E. Pawson and T. Brooking, 169–82. Melbourne: Oxford University Press.

Brown, David E., and Charles H. Lowe. 1980. *Biotic Communities of the Southwest*. General Technical Report RM-78. Fort Collins, CO: USDA Forest Service, Rocky Mountain Forest and Range Experiment Station.

Buckingham, William A. 1982. *Operation Ranch Hand: The Air Force and Herbicides in Southeast Asia, 1961–1971*. Washington, DC: Office of Air Force History, United States Air Force.

Buffington, Lee C., and Carlton H. Herbel. 1965. "Vegetational Changes on a Semidesert Grassland Range from 1858 to 1963." *Ecological Monographs* 35:140–64.

Cable, Dwight R. 1965. "Damage to Mesquite, Lehmann Lovegrass, and Black Grama by a Hot June Fire." *Journal of Range Management* 18:326–29.

———. 1967. "Fire Effects on Semidesert Grasses and Shrubs." *Journal of Range Management* 20:170–76.

———. 1971. "Lehmann Lovegrass on the Santa Rita Experimental Range, 1937–1968." *Journal of Range Management* 24:17–21.

Cable, Dwight R., and Fred H. Tschirley. 1961. "Responses of Native and Introduced Grasses Following Aerial Spraying of Velvet Mesquite in Southern Arizona." *Journal of Range Management* 14:155–59.

Calef, Wesley. 1960. *Private Grazing and Public Lands: Studies of the Local Management of the Taylor Grazing Act.* Chicago: University of Chicago Press.

Cameron, Jenks. 1929. *The Bureau of Biological Survey: Its History, Activities and Organization.* Institute for Government Research [Brookings Institution] Service Monographs of the United States Government, no. 54. Baltimore: Johns Hopkins Press.

Campbell, R. S. 1931. "Plant Succession and Grazing Capacity on Clay Soils in Southern New Mexico." *Journal of Agricultural Research* 43:1027–51.

———. 1948. "Milestones in Range Management." *Journal of Range Management* 1:4–8.

Canfield, R. H. 1941. "Application of the Line Interception Method in Sampling Range Vegetation." *Journal of Forestry* 39:388–94.

Carle, David. 2002. *Burning Questions: America's Fight with Nature's Fire.* Westport, CT: Praeger.

Cawelti, John G. 1984. *The Six-Gun Mystique.* 2nd ed. Bowling Green, OH: Bowling Green University Popular Press.

Chambers, Jeanne, Helena Deswood, Emile Elias, Kris Havstad, Amber Kerr, Albert Rango, Mark Schwartz, Kerri Steenwerth, Caiti Steele, and Peter Stine. 2015. *Southwest Regional Climate Hub and California Subsidiary Hub Assessment of Climate Change Vulnerability and Adaptation and Mitigation Strategies.* Washington, DC: USDA Climate Hub Program.

Chapline, W. R. 1921. "Report of Office of Grazing Studies." Typescript located in the Records of the Forest Service (Record Group 95), National Archives and Records Administration, College Park, MD.

———. 1925. "Range Investigations." Annual Report. Typescript located in the Records of the Forest Service (Record Group 95), National Archives and Records Administration, College Park, MD.

———. 1928. "Range Research." Annual Report. Typescript located in the Records of the Forest Service (Record Group 95), National Archives and Records Administration, College Park, MD.

———. 1933. "Range Research Methods of the U.S. Forest Service." Dated July 27, labeled as "Presented at the World's Grain Conference, August 1, Regina, Saskatchewan, Canada." Typescript located in the Records of the Forest Service (Record Group 95), National Archives and Records Administration, College Park, MD.

———. 1937. "Range Research in the United States." *Herbage Reviews* 5:1–13.

———. 1948. "Grazing on Range Lands." In *Yearbook of the Department of Agriculture,* 212–16. Washington, DC: USDA.

———. 1951. "Range Management History and Philosophy." *Journal of Forestry* 49:634–38.

———. 1980. "First 10 Years of the Office of Grazing Studies." *Rangelands* 2:223–27.

Chapline, W. R., R. S. Campbell, R. Price, and G. Stewart. 1944. "The History of Western Range Research." *Agricultural History* 18:127–43.

Chapman, Royal N. 1928. "The Quantitative Analysis of Environmental Factors." *Ecology* 9:111–22.

Charnley, Susan, Thomas E. Sheridan, and Gary P. Nabhan. 2014. *Stitching the West Back Together: Conservation of Working Landscapes*. Chicago: University of Chicago Press.

Clapp, Earle H. 1926. *A National Program of Forest Research*. Washington, DC: American Tree Association for the Society of American Foresters.

Clawson, Marion. 1948. "Range Forage Conditions in Relation to Annual Precipitation." *Land Economics* 24:264–80.

Clements, Edith S. 1960. *Adventures in Ecology; Half a Million Miles: From Mud to Macadam*. New York: Pageant Press.

Clements, Frederic E. 1904. *The Development and Structure of Vegetation*. Lincoln: University of Nebraska Botanical Seminar.

———. 1916. *Plant Succession: An Analysis of the Development of Vegetation*. Publication no. 242. Washington, DC: Carnegie Institution of Washington.

———. 1920. *Plant Indicators: The Relation of Plant Communities to Process and Practice*. Publication no. 290. Washington, DC: Carnegie Institution of Washington.

———. 1934. "The Relict Method in Dynamic Ecology." *Journal of Ecology* 22:39–68.

Coase, R. H. 1960. "The Problem of Social Cost." *Journal of Law and Economics* 3:1.

Cohn, Theodore. 1975. "The Sahelian Drought: Problems of Land Use." *International Journal* 30:428–44.

Coldwell, Thomas, and Joseph F. Pechanec. 1950. "The Brushland Plow." *Research Notes* 64. USDA Pacific Northwest Forest and Range Experiment Station.

Coleman, David C. 2010. *Big Ecology: The Emergence of Ecosystem Science*. Berkeley: University of California Press.

Conley, Marsha Reeves, and Walt Conley. 1984. *New Mexico State University College Ranch and Jornada Experimental Range: A Summary of Research, 1900–1983*. Special Report no. 56. Las Cruces: New Mexico State University Agricultural Experiment Station.

Connelly, Matthew James. 2008. *Fatal Misconception: The Struggle to Control World Population*. Cambridge, MA: Belknap Press of Harvard University Press.

Corbett, Jim. 2005. *Sanctuary for All Life: The Cowballah of Jim Corbett*. Berthoud, CO: Howling Dog Press.

Cottam, Walter P., and Frederick R. Evans. 1945. "A Comparative Study of the Vegetation of Grazed and Ungrazed Canyons of the Wasatch Range, Utah." *Ecology* 26:171–81.

Cotton, J. S. 1904. *A Report on the Range Conditions of Central Washington*. Bulletin no. 60. Pullman: Washington State Agricultural College and School of Science, Experiment Station.

Coville, Frederick Vernon. 1898. *Forest Growth and Sheep Grazing in the Cascade Mountains of Oregon*. Bulletin no. 15. Washington, DC: USDA Division of Forestry.

Cowden, Ray. 1958. "Brush Land Control." In *Your Range—Its Management*. Special Report 2, 14–15. Tucson: Arizona Agricultural Extension Service.

Cox, Jerry R., and Gilbert L. Jordan. 1983. "Density and Production of Seeded Range Grasses in Southeastern Arizona (1970–1982)." *Journal of Range Management* 36:649–52.

Cox, J. R., H. L. Morton, T. N. Johnsen Jr., G. L. Jordan, S. C. Martin, and L. C. Fierro. 1982. *Vegetation Restoration in the Chihuahuan and Sonoran Deserts of North America*. Agricultural Reviews and Manuals ARM-W-28. Washington, DC: USDA Agricultural Research Service.

Cox, Kevin R. 1998. "Spaces of Dependence, Spaces of Engagement and the Politics of Scale, or: Looking for Local Politics." *Political Geography* 17:1–23.

Crider, Franklin J. 1945. *Three Introduced Lovegrasses for Soil Conservation*. Circular no. 730. Washington, DC: USDA.

Crosby, Alfred W. 1986. *Ecological Imperialism: The Biological Expansion of Europe, 900–1900*. Cambridge: Cambridge University Press.

Culhane, Paul J. 1981. *Public Lands Politics: Interest Group Influence on the Forest Service and the Bureau of Land Management*. Baltimore: Johns Hopkins University Press.

Culley, M. J., and R. R. Hill. 1926. "Manuscript Covering Range and Stock Investigations on the Santa Rita Range Reserve." Typescript located in the Records of the Forest Service (Record Group 95), National Archives and Records Administration, College Park, MD.

Dahl, Gudrun. 1979. *Suffering Grass: Subsistence and Society of Waso Borana*. Stockholm Studies in Social Anthropology 8. Stockholm: Department of Social Anthropology, University of Stockholm.

Dahl, Gudrun, and Anders Hjort. 1976. *Having Herds: Pastoral Herd Growth and Household Economy*. Stockholm Studies in Social Anthropology 2. Stockholm: Department of Social Anthropology, University of Stockholm.

Davis, Diana K. 2007. *Resurrecting the Granary of Rome: Environmental History and French Colonial Expansion in North Africa*. Athens: Ohio University Press.

———. 2016. *The Arid Lands: History, Power, Knowledge*. Cambridge, MA: MIT University Press.

Davis, Richard C., ed. 1983. *Encyclopedia of American Forest and Conservation History*. New York: Macmillan.

DeLay, Brian. 2008. *War of a Thousand Deserts: Indian Raids and the U.S.-Mexican War*. New Haven: Yale University Press.

DeLuca, Kevin, and Anne Demo. 2001. "Imagining Nature and Erasing Class and Race: Carleton Watkins, John Muir, and the Construction of Wilderness." *Environmental History* 6:541–60.

Dillman, A. C. 1946. "The Beginnings of Crested Wheatgrass in North America." *Journal of the American Society of Agronomy* 38:237–50.

Dobyns, Henry F. 1981. *From Fire to Flood: Historic Human Destruction of Sonoran Desert Riverine Oases*. Socorro, NM: Ballena Press.

Donahue, Debra L. 1999. *The Western Range Revisited: Removing Livestock from Public Lands to Conserve Native Biodiversity*. Norman: Oklahoma University Press.

Dunbar, Gary S. 1970. "African Ranches Ltd., 1914–1931: An Ill-Fated Stockraising Enterprise in Northern Nigeria." *Annals of the Association of American Geographers* 60:102–23.

Dwyer, Don D., and Rex D. Pieper. 1967. "Fire Effects on Blue Grama-Pinyon-Juniper Rangeland in New Mexico." *Journal of Range Management* 20:359–62.

Dyksterhuis, E. J. 1949. "Condition and Management of Range Land Based on Quantitative Ecology." *Journal of Range Management* 2:104–15.

Eckholm, Erik P. 1975. "Desertification: A World Problem." *Ambio* 4:137–45.

Edge, Rosalie, and the Emergency Conservation Committee. 1934. *The United States Bureau of Destruction and Extermination, the Misnamed and Perverted "Biological Survey."* Emergency Conservation Committee.

Egan, Timothy. 2009. *The Big Burn: Teddy Roosevelt and the Fire That Saved America*. Boston: Houghton Mifflin Harcourt.

Ellis, James E., and David M. Swift. 1988. "Stability of African Pastoral Ecosystems: Alternate Paradigms and Implications for Development." *Journal of Range Management* 41:450–59.

Ellison, Lincoln. 1949. "The Ecological Basis for Judging Condition and Trend on Mountain Range Land." *Journal of Forestry* 47:786–95.

———. 1960. "Influence of Grazing on Plant Succession of Rangelands." *Botanical Review* 26:1–78.

Ferguson, James. 1990. *The Anti-Politics Machine: "Development," Depoliticization, and Bureaucratic Power in Lesotho*. Cambridge: Cambridge University Press.

Flory, Evan L., and Charles G. Marshall. 1942. *Regrassing for Soil Protection in the Southwest*. Farmers' Bulletin no. 1913. Washington, DC: USDA.

Food and Agriculture Organization of the United Nations (FAO). 1967. *East African Livestock Survey; Regional: Kenya, Tanzania, Uganda*. Rome: United Nations Development Program [and] Food and Agriculture Organization of the United Nations.

Forsling, Clarence Luther. 1951. "Relation of Sustained Livestock Production to Condition of Grazing Land." In *Proceedings*, Vol. 6. United Nations Scientific Conference on the Conservation and Utilization of Resources, 500–505. Lake Success, NY: United Nations Department of Economic Affairs.

Forsling, Clarence Luther, and William Adams Dayton. 1931. *Artificial Reseeding on Western Mountain Range Lands*. Circular 178. Washington, DC: USDA.

Forstater, Mathew. 2002. "Bones for Sale: 'Development,' Environment and Food Security in East Africa." *Review of Political Economy* 14:47–67.

Forsyth, Tim. 2003. *Critical Political Ecology: The Politics of Environmental Science*. London: Routledge.

Foss, Phillip O. 1960a. *The Grazing Fee Dilemma*. Birmingham: Published for the Inter-University Case Program by University of Alabama Press.

———. 1960b. *Politics and Grass: The Administration of Grazing on the Public Domain*. Seattle: University of Washington Press.

Fowler, John M., and James R. Gray. 1988. "Rangeland Economics in the Arid West." In *Rangelands*, edited by Bruce A. Buchanan, 67–89. Albuquerque: University of New Mexico Press.

Fratkin, Elliot. 1997. "Pastoralism: Governance and Development Issues." *Annual Review of Anthropology*, 235–61.

Frederick, Aaron W. 1910. "Letter to the Editor." *American Forestry* 16:563.

Friedel, M. H. 1991. "Range Condition Assessment and the Concept of Thresholds: A Viewpoint." *Journal of Range Management* 44:422–26.

Frink, Maurice, W. Turrentine Jackson, and Agnes Wright Spring. 1956. *When Grass Was King*. Boulder: University of Colorado Press.

Fuhlendorf, Samuel D., David D. Briske, and Fred E. Smeins. "Herbaceous Vegetation Change In Variable Rangeland Environments: The Relative Contribution of Grazing and Climatic Variability." *Applied Vegetation Science* 4, no. 2 (2001): 177–88.

Fuhlendorf, Samuel D., and David M. Engle. 2001. "Restoring Heterogeneity on Rangelands: Ecosystem Management Based on Evolutionary Grazing Patterns." *BioScience* 51: 625–32.

Fuhlendorf, Samuel D., and Fred E. Smeins. 1999. "Scaling Effects of Grazing in a Semi-Arid Grassland." *Journal of Vegetation Science* 10: 731–38.

Fuhlendorf, Samuel D., Fred E. Smeins, and William E. Grant. "Simulation of a Fire-sensitive Ecological Threshold: A Case Study of Ashe Juniper on the Edwards Plateau of Texas, USA." *Ecological Modelling* 90, no. 3 (1996): 245–55.

Galvin, Kathleen A., Robin S. Reid, R. H. Behnke, and N. Thompson Hobbs, eds. 2008. *Fragmentation in Semi-Arid and Arid Landscapes: Consequences for Human and Natural Systems*. Dordrecht: Springer.

Gardner, J. L. 1950. "Effects of Thirty Years of Protection from Grazing in Desert Grassland." *Ecology* 31:44–50.

———. 1963. "Aridity and Agriculture." In *Aridity and Man*, edited by C. Hodge and P. C. Duisberg, 239–76. Washington, DC: American Association for the Advancement of Science.

Gentner, Bradley J., and John A. Tanaka. 2002. "Classifying Federal Public Land Grazing Permittees." *Journal of Range Management* 55:2–11.

Gibbens, R. P., R. F. Beck, R. P. McNeely, and C. H. Herbel. 1992. "Recent Rates of Mesquite Establishment in the Northern Chihuahuan Desert." *Journal of Range Management* 45:585–88.

Gilles, Jere Lee, and C. de Haan. 1994. *Recent Trends in World Bank Pastoral Development Projects: A Review of 13 Bank Projects in Light of "the New Pastoral Ecology."* Pastoral Development Network Paper. London: Overseas Development Institute.

Gleason, Henry A. 1926. "The Individualistic Concept of the Plant Association." *Bulletin of the Torrey Botanical Club* 53:7–26.

———. 1939. "The Individualistic Concept of the Plant Association." *American Midland Naturalist* 21:92–110.

Glendening, George E. 1942. "Germination and Emergence of Some Native Grasses in Relation to Litter Cover and Soil Moisture." *Agronomy Journal* 34:797–804.

———. 1952. "Some Quantitative Data on the Increase of Mesquite and Cactus on a Desert Grassland Range in Southern Arizona." *Ecology* 33:319–28.

Glendening, George E., and Kenneth W. Parker. 1950. "Problem Analysis: 'Range Reseeding in the Southwest.'" Originally dated 1948, revised in May 1950 by H. G. Reynolds. Typescript located in the Records of the Forest Service (Record Group 95), National Archives and Records Administration, College Park, MD.

Glendening, George E., and Harold A. Paulsen. 1955. *Reproduction and Establishment of Velvet Mesquite: As Related to Invasion of Semidesert Grasslands.* Technical Bulletin sno. 1127. Washington, DC: USDA.

Godfrey, Anthony. 2013. *The Search for Forest Facts: A History of the Pacific Southwest Forest and Range Experiment Station, 1926–2000.* General Technical Report PSW-GTR-233. USDA Forest Service, Pacific Southwest Research Station.

Goodloe, Sid. 1969. "Short Duration Grazing in Rhodesia." *Journal of Range Management* 22:369–73.

Gori, David F., and Carolyn A. F. Enquist. 2003. *An Assessment of the Spatial Extent and Condition of Grasslands in Central and Southern Arizona, Southwestern New Mexico and Northern Mexico.* Prepared by the Nature Conservancy, Arizona Chapter.

Government Accounting Office. 1991. *Rangeland Management: Forest Service Not Performing Needed Monitoring of Grazing Allotments.* GAO/RCED-91-148. Washington, DC: GAO.

Graves, Henry Solon. 1910. *Protection of Forests from Fire.* Washington, DC: USDA Forest Service.

Grice, A. C., and K. C. Hodgkinson. 2002. *Global Rangelands: Progress and Prospects.* Wallingford, UK: Centre for Agriculture and Bioscience International.

Griffin, G. F., and M. H. Friedel. 1984. "Effects of Fire on Central Australian Rangelands. II. Changes in Tree and Shrub Populations." *Australian Journal of Ecology* 9:395–403.

———. 1985. "Discontinuous Change in Central Australia: Some Implications of Major Ecological Events for Land Management." *Journal of Arid Environments* 9:63–80.

Griffin, G. F., N. F. Price, and H. F. Portlock. 1983. "Wildfires in the Central Australian Rangelands, 1970–1980." *Journal of Environmental Management* 17:311–23.

Griffiths, David. 1904. *Range Investigations in Arizona.* USDA Bureau of Plant Industry Bulletin 67. Washington, DC: Government Printing Office.

———. 1907. *The Reseeding of Depleted Range and Native Pastures.* USDA Bureau of Plant Industry Bulletin 117. Washington, DC: Government Printing Office.

———. 1910. *A Protected Stock Range in Arizona.* USDA Bureau of Plant Industry Bulletin 177. Washington, DC: Government Printing Office.

Griffiths, Tom, and Libby Robin, eds. 1997. *Ecology and Empire: Environmental History of Settler Societies.* Seattle: University of Washington Press.

Grove, Alfred T. 1974. "Desertification in the African Environment." *African Affairs* 73:137–51.

Hadley, R. F., B. Dasgupta, N. E. Reynolds, S. Sandford, and E. B. Worthington. 1977. "Evaluation of Land-Use and Land-Treatment Practices in Semi-Arid Western United States [and Discussion]." *Philosophical Transactions of the Royal Society B: Biological Sciences* 278:543–54.

Hagen, Joel B. 1993. "Clementsian Ecologists: The Internal Dynamics of a Research School." *Osiris* 8:178–95.

Hansen, Andrew J., Richard L. Knight, John M. Marzluff, Scott Powell, Kathryn Brown, Patricia H. Gude, and Kingsford Jones. 2005. "Effects of Exurban Development on Biodiversity: Patterns, Mechanisms, and Research Needs." *Ecological Applications* 15:1893–1905.

Hardin, Garrett. 1968. "The Tragedy of the Commons." *Science* 162:1243–48.

Hatton, John H. 1920. *Live-Stock Grazing as a Factor in Fire Protection on the National Forests.* Circular 134. Washington, DC: USDA.

Havstad, Kris M., Laura F. Huenneke, and William H. Schlesinger, eds. 2006. *Structure and Function of a Chihuahuan Desert Ecosystem: The Jornada Basin Long-Term Ecological Research Site.* New York: Oxford University Press.

Havstad, Kris M., Debra P. C. Peters, Rhonda Skaggs, Joel Brown, Brandon Bestelmeyer, Ed Fredrickson, Jeffrey Herrick, and Jack Wright. 2007. "Ecological Services to and from Rangelands of the United States." *Ecological Economics* 64:261–68.

Havstad, K. M., and J. E. Herrick. 2003. "Long-Term Ecological Monitoring." *Arid Land Research and Management* 17:389–400.

Heady, Harold F. 1960. *Range Management in East Africa.* Nairobi: Government Printer.

Herrick, J. E., K. M. Havstad, and D. P. Coffin. 1996. "Rethinking Remediation Technologies for Desertified Landscapes." *Journal of Soil and Water Conservation* 52:220.

Herskovits, Melville J. 1926. "The Cattle Complex of East Africa. Part One." *American Anthropologist* 28:230–72.

Heyboer, Maarten. 1992. "Grass-Counters, Stock-Feeders, and the Dual Orientation of Applied Science: The History of Range Science, 1895–1960." PhD dissertation, Virginia Polytechnic Institute and State University.

Hirt, Paul W. 1994. *A Conspiracy of Optimism: Management of the National Forests since World War Two.* Lincoln: University of Nebraska Press.

Holechek, Jerry. 1981. "A Brief History of Range Management in the United States." *Rangelands* 3:16–18.

Holechek, Jerry, Rex D. Pieper, and Carlton H. Herbel. 2004. *Range Management: Principles and Practices.* 5th ed. Upper Saddle River, NJ: Pearson/Prentice Hall.

Holland, Peter, Kevin O'Connor, and Alexander Wearing. 2002. "Remaking the Grasslands of the Open Country." In *Environmental Histories of New Zealand,* edited by E. Pawson and T. Brooking, 69–83. Melbourne: Oxford University Press.

Holling, C. S. 1973. "Resilience and Stability of Ecological Systems." *Annual Review of Ecology and Systematics,* 1–23.

———. 1978. *Adaptive Environmental Assessment and Management.* New York: Wiley.

Holling, C. S., and Gary K. Meffe. 1996. "Command and Control and the Pathology of Natural Resource Management." *Conservation Biology* 10:328–37.

Hoover, M. M. 1939. *Native and Adapted Grasses for Conservation of Soil and Moisture in the Great Plains and Western States.* Farmers' Bulletin 1912. Washington, DC: USDA.

Hornby, H. E. 1936. "Overstocking in Tanganyika Territory." *East African Agricultural Journal* 1:353–60.

Hull, A. C. Jr., and G. J. Klomp. 1966. "Longevity of Crested Wheatgrass in the Sagebrush-Grass Type in Southern Idaho." *Journal of Range Management* 19:5–11.

Humphrey, Robert R. 1949. "Fire as a Means of Controlling Velvet Mesquite, Burroweed, and Cholla on Southern Arizona Ranges." *Journal of Range Management* 2:175–82.

———. 1958. "The Desert Grassland: A History of Vegetational Change and an Analysis of Causes." *Botanical Review* 24:193–252.

Hutchinson, Charles F. 1996. "The Sahelian Desertification Debate: A View from the American South-West." *Journal of Arid Environments* 33:519–24.

Hyder, D. N., R. E. Bement, E. E. Remmenga, and C. Terwilliger Jr. 1966. "Vegetation-Soils and Vegetation-Grazing Relations from Frequency Data." *Journal of Range Management* 19:11–17.

Hyder, Donald N. 1954. *Spray to Control Big Sagebrush.* Corvallis: Oregon State College Agricultural Experiment Station.

Illius, A. W., and T. G. O'Connor. 1999. "On the Relevance of Nonequilibrium Concepts to Arid and Semiarid Grazing Systems." *Ecological Applications* 9:798–813.

Ingold, Tim. 1980. *Hunters, Pastoralists, and Ranchers: Reindeer Economies and Their Transformations.* Cambridge: Cambridge University Press.

Inter-Agency Range Survey Committee. 1937. *Instructions for Range Surveys.* Washington, DC: Inter-Agency Range Survey Committee.

Inter-American Conference on Conservation of Renewable Natural Resources (IACCRNR). 1948. *Proceedings of the Inter-American Conference on Conservation of Renewable Natural Resources: Denver, Colorado, September 7–20, 1948.* Publication 3382. Washington, DC: Department of State, Office of Public Affairs.

Isenberg, Andrew C. 2000. *The Destruction of the Bison: An Environmental History, 1750–1920.* Cambridge: Cambridge University Press.

———. 2014. "Seas of Grass: Grasslands in World Environmental History." In *Oxford Handbook of Environmental History,* edited by Andrew C. Isenberg, 133–53. Cary, NC: Oxford University Press.

James, Edwin, comp. 1823. *Account of an Expedition from Pittsburgh to the Rocky Mountains: Performed in the Years 1819 and '20, by Order of the Hon. JC Calhoun, Sec'y of War: Under the Command of Major Stephen H. Long.* 2 vols. Philadelphia: H. C. Carey and I. Lea.

Jardine, James T. 1909. *Coyote-Proof Pasture Experiment, 1908.* Circular no. 160. Washington, DC: USDA Forest Service.

———. 1910. *The Pasturage System for Handling Range Sheep: Investigations during 1909.* Circular no. 178. Washington, DC: USDA Forest Service.

———. 1911. *Coyote-Proof Inclosures in Connection with Range Lambing Grounds.* USDA Forest Service Bulletin no. 97. Washington, DC: Government Printing Office.

———. 1915. *Grazing.* Manuscript. New Haven, CT: Yale Forestry School.

———. 1916. Annual Report of the Inspector of Grazing. Records of the Forest Service (Record Group 95), National Archives and Records Administration, College Park, MD.

———. 1917. "Report of Grazing Studies." Annual Report. Records of the Forest Service (Record Group 95), National Archives and Records Administration, College Park, MD.

———. 1919. "Range Investigations." Annual Report of the Inspector of Grazing. Records of the Forest Service (Record Group 95), National Archives and Records Administration, College Park, MD.

Jardine, James T., and Mark Anderson. 1919. *Range Management on the National Forests*. Bulletin no. 790. Washington, DC: USDA.

Jardine, James T., and Frederick V. Coville. 1908. *Preliminary Report on Grazing Experiments in a Coyote-Proof Pasture*. Circular no. 156. Washington, DC: USDA Forest Service.

Jardine, James T., and Leon C. Hurtt. 1917. *Increased Cattle Production on Southwestern Ranges*. Bulletin no. 588. Washington, DC: USDA.

Jordan, Terry G. 1993. *North American Cattle-Ranching Frontiers: Origins, Diffusion, and Differentiation*. Albuquerque: University of New Mexico Press.

Joss, P. J., P. W. Lynch, and O. B. Williams, eds. 1986. *Rangelands: A Resource under Siege: Proceedings of the Second International Rangeland Congress*. Cambridge: Cambridge University Press.

Joyce, Linda A. 1993. "The Life Cycle of the Range Condition Concept." *Journal of Range Management* 46:132–38.

Kaufert, F. H., and W. H. Cummings. 1955. *Forestry and Related Research in North America*. Washington, DC: Society of American Foresters.

Kinsey, Joni L., Rebecca Roberts, and Robert F. Sayre. 1996. "Prairie Prospects: The Aesthetics of Plainness." *Prospects* 21:261–97.

Kloppenburg, Jack Ralph. 1988. *First the Seed: The Political Economy of Plant Biotechnology, 1492–2000*. Cambridge: Cambridge University Press.

Knapp, Corrine Noel and Maria Fernández-Giménez. 2008. "Knowing the Land: A Review of Local Knowledge Revealed in Ranch Memoirs." *Rangeland Ecology and Management* 61:148–55.

———. 2009a. "Knowledge in Practice: Documenting Rancher Local Knowledge in Northwest Colorado." *Rangeland Ecology and Management* 62:500–509.

———. 2009b. "Understanding Change: Integrating Rancher Knowledge into State-and-Transition Models." *Rangeland Ecology and Management* 62:510–21.

Knapp, Paul A. 1996. "Cheatgrass (*Bromus Tectorum* L) Dominance in the Great Basin Desert: History, Persistence, and Influences to Human Activities." *Global Environmental Change* 6:37–52.

Koford, Carl B. 1958. "Prairie Dogs, Whitefaces, and Blue Grama." *Wildlife Monographs* 3:3–78.

Korstian, Clarence F. 1921. "Grazing Practice on the National Forests and Its Effect on Natural Conditions." *Scientific Monthly* 13:275–81.

Kosek, Jake. 2006. *Understories: The Political Life of Forests in Northern New Mexico*. Durham, NC: Duke University Press.

Kwa, Chunglin. 1987. "Representations of Nature Mediating between Ecology and Science Policy: The Case of the International Biological Programme." *Social Studies of Science* 17:413–42.

———. 1993. "Modeling the Grasslands." *Historical Studies in the Physical and Biological Sciences* 24:125–55.

Langston, Nancy. 1995. *Forest Dreams, Forest Nightmares: The Paradox of Old Growth in the Inland West*. Seattle: University of Washington Press.

Lantow, J. L., and E. L. Flory. 1940. "Fluctuating Forage Production. Its Significance in Proper Range and Livestock Management on Southwestern Ranges." *Soil Conservation* 6:137–44.

Lantz, David E. 1905. *Coyotes in Their Economic Relations*. USDA Biological Survey Bulletin no. 20. Washington, DC: Government Printing Office.

———. 1918. "The House Rat: The Most Destructive Animal in the World." In *Yearbook of the Department of Agriculture 1917*, 235–51. Washington, DC: Government Printing Office.

Larson, J. E. 1980. *Revegetation Equipment Catalog*. Missoula, MT: USDA Forest Service Equipment Development Center.

Laycock, W. A. 1991. "Stable States and Thresholds of Range Condition on North American Rangelands: A Viewpoint." *Journal of Range Management* 44:427–33.

Lebel, Louis, Po Garden, and Masao Imamura. 2005. "The Politics of Scale, Position, and Place in the Governance of Water Resources in the Mekong Region." *Ecology and Society* 10(2):18 (online).

Le Houerou, Henri Noel. 1980. "The Rangelands of the Sahel." *Journal of Range Management* 33:41–46.

Leopold, Aldo. 1924. "Grass, Brush, Timber, and Fire in Southern Arizona." *Journal of Forestry* 22:1–10.

———. 1970. *A Sand County Almanac, with Essays on Conservation from Round River*. New York: Ballantine Books.

———. 1992. "Ecology and Politics." In *The River of the Mother of God and Other Essays by Aldo Leopold*, edited by Susan L. Flader and J. Baird Callicott, 281–86. Madison: University of Wisconsin Press.

Levin, Simon A. 1992. "The Problem of Pattern and Scale in Ecology: The Robert H. MacArthur Award Lecture." *Ecology* 73:1943–67.

Li, Wenjun, and Lynn Huntsinger. 2011. "China's Grassland Contract Policy and Its Impacts on Herder Ability to Benefit in Inner Mongolia: Tragic Feedbacks." *Ecology and Society* 16(2):1 (online).

Limerick, Patricia Nelson. 1987. *The Legacy of Conquest: The Unbroken Past of the American West*. New York: Norton.

Livneh, B., E. A. Rosenberg, C. Lin, B. Nijssen, V. Mishra, K. M. Andreadis, E. P. Maurer, and D. P. Lettenmaier. 2013. "A Long-Term Hydrologically Based Dataset of Land Surface Fluxes and States for the Conterminous United States: Update and Extensions. *Journal of Climate* 26:9384–92.

Lofchie, Michael F. 1975. "Political and Economic Origins of African Hunger." *Journal of Modern African Studies* 13:551–67.

Lynam, T. J. P., and M. Stafford Smith. 2004. "Monitoring in a Complex World—Seeking Slow Variables, a Scaled Focus, and Speedier Learning." *African Journal of Range and Forage Science* 21:69–78.

MacDonald, Glen M. 2010. "Water, Climate Change, and Sustainability in the Southwest." *Proceedings of the National Academy of Sciences* 107:21256–62.

Maher, Neil M. 2008. *Nature's New Deal: The Civilian Conservation Corps and the Roots of the American Environmental Movement*. Oxford: Oxford University Press.

Malin, James C. 1956. *The Grasslands of North America: Prolegomena to Its History with Addenda*. Lawrence, KS: James C. Malin.

Mann, Charles C. 2006. *1491: New Revelations of the Americas before Columbus*. New York: Vintage.

Markakis, John. 2004. *Pastoralism on the Margin*. London: Minority Rights Group International.

Marris, Emma. 2014. "Rethinking Predators: Legend of the Wolf." *Nature* 507:158–60.

Martin, S. Clark. 1966. *The Santa Rita Experimental Range: A Center for Research on Improvement and Management of Semidesert Rangelands*. Research Paper RM-22. Fort Collins, CO: USDA Forest Service, Rocky Mountain Forest and Range Experiment Station.

———. 1973. "Responses of Semidesert Grasses to Seasonal Rest." *Journal of Range Management* 26:165–70.

———. 1978. "The Santa Rita Grazing System." In *Proceedings of the First International Rangeland Congress,* edited by D. N. Hyder, 573–75. Denver, CO: Society for Range Management.

———. 1983. "Responses of Semidesert Grasses and Shrubs to Fall Burning." *Journal of Range Management* 36:604–10.

Martin, S. Clark, and Dwight R. Cable. 1975. "Highlights of Research on the Santa Rita Experimental Range." In *Arid Shrublands: Proceedings of the Third Workshop of the United States/Australia Rangelands Panel, Tucson, Arizona,* edited by D. N. Hyder, 51–57. Denver, CO: Society for Range Management.

Martin, S. Clark, and Hudson G. Reynolds. 1973. "The Santa Rita Experimental Range: Your Facility for Research on Semidesert Ecosystems." *Journal of the Arizona Academy of Science* 8:56–67.

Maxon, William R. 1937. "Frederick Vernon Coville." *Science* 85:280–81.

May, Robert M. 1977. "Thresholds and Breakpoints in Ecosystems with a Multiplicity of Stable States." *Nature* 269:471–77.

McClaran, Mitchel P., Peter F. Ffolliott, and Carleton B. Edminster, eds. 2003. *Santa Rita Experimental Range: 100 Years (1903–2003) of Accomplishments and Contributions.* Proc. RMRS-P-30. Ogden, UT: USDA Rocky Mountain Research Station.

McGinnies, William G., Kenneth W. Parker, and George E. Glendening. 1941. *Southwestern Range Ecology*. Washington, DC: US Forest Service.

McPherson, Guy R., and Jake F. Weltzin. 2000. *Disturbance and Climate Change in United States/Mexico Borderland Plant Communities: A State-of-the-Knowledge Review.* General Technical Report RMRS GTR-50. Fort Collins, CO: USDA Forest Service, Rocky Mountain Research Station.

Meine, Curt. 1988. *Aldo Leopold: His Life and Work.* Madison: University of Wisconsin Press.

Merriam, C. H. 1902. "The Prairie Dog of the Great Plains." In *Yearbook of the Department of Agriculture 1901*, 257–70. Washington, DC: USDA.

Merrill, Karen R. 2002. *Public Lands and Political Meaning: Ranchers, the Government, and the Property between Them.* Berkeley: University of California Press.

Meuret, Michel, and Fred Provenza, eds. 2014. *The Art and Science of Shepherding: Tapping the Wisdom of French Herders.* Translated by Bruce Inksetter and Melanie Guedenet. Austin, TX: Acres USA.

Milchunas, Daniel G., and William K. Lauenroth. 1993. "Quantitative Effects of Grazing on Vegetation and Soils over a Global Range of Environments." *Ecological Monographs* 63:327–66.

Millennium Ecosystem Assessment. 2005. *Ecosystems and Human Well-Being*. Vol. 5. Washington, DC: Island Press.

Muhn, James Allan. 1987. "The Mizpah-Pumpkin Creek Grazing District: Its History and Influence on the Enactment of a Public Lands Grazing Policy, 1926–1934." MA thesis, Montana State University.

Munoz-Erickson, Tischa A., Bernardo Aguilar-Gonzalez, and Thomas D. Sisk. 2007. "Linking Ecosystem Health Indicators and Collaborative Management: A Systematic Framework to Evaluate Ecological and Social Outcomes." *Ecology and Society* 12:6.

National Research Council. 1994. *Rangeland Health: New Methods to Classify, Inventory, and Monitor Rangelands.* Washington, DC: National Academy Press.

Nelson, Enoch W. 1934. "The Influence of Precipitation and Grazing upon Black Grama Grass Range." Technical Bulletin 409. Washington, DC: US Department of Agriculture.

Netz, Reviel. 2004. *Barbed Wire: An Ecology of Modernity.* Middletown, CT: Wesleyan University Press.

Noy-Meir, Imanuel. 1973. "Desert Ecosystems: Environment and Producers." *Annual Review of Ecology and Systematics*, 25–51.

———. 1975. "Stability of Grazing Systems: An Application of Predator-Prey Graphs." *Journal of Ecology* 63:459–81.

———. 1979. "Structure and Function of Desert Ecosystems." *Israel Journal of Botany* 28:1–19.

Oakes, Claudia Lea. 2000. "History and Consequence of Keystone Mammal Eradication in the Desert Grasslands: The Arizona Black-Tailed Prairie Dog (*Cynomys Ludovicianus Arizonensis*)." PhD dissertation, University of Texas at Austin.

Olberding, Susan Deaver, John E. Mitchell, and Margaret M. Moore. 2005. "'Doing the Best We Could with What We Had': USFS Range Research in the Southwest." *Rangelands* 27:29–36.

Oldemeyer, John L., Dean E. Biggins, and Brian J. Miller, eds. 1993. *Proceedings of the Symposium on the Management of Prairie Dog Complexes for the Reintroduction of the Black-Footed Ferret.* Washington, DC: US Department of the Interior, Fish and Wildlife Service.

Osborn, Fairfield. 1950. "The World Resources Situation." In *Proceedings*, Vol. 1. United Nations Scientific Conference on the Conservation and Utilization of Resources, 12–15. Lake Success, NY: United Nations Department of Economic Affairs.

Osgood, Ernest Staples. 1929. *The Day of the Cattleman.* Minneapolis: University of Minnesota Press.

Osgood, Wilfred H. 1944. "Biographical Memoir of Clinton Hart Merriam, 1855–1942. *National Academy of Sciences Biographical Memoirs* 24:1–57.

Owensby, Clenton E. 2000. "Achievements in Management and Utilization of Privately-Owned Rangelands." *Journal of Range Management* 53:12–16.

Palmer, Theodore S. 1897a. "Extermination of Noxious Animals by Bounties." In *Yearbook of the Department of Agriculture 1896*, 55–68. Washington, DC: USDA.

———. 1897b. *The Jack Rabbits of the United States.* Division of Biological Survey Bulletin no. 8. Washington, DC: USDA.

Parker, Kenneth W. 1938. *Effect of Jackrabbits on the Rate of Recovery of Deteriorated Rangeland.* Press Bulletin no. 839. Albuquerque: New Mexico State University Agricultural Experiment Station.

———. 1939. *The Control of Burroweed.* Research Note 72. USDA Forest Service, Southwestern Forest and Range Experiment Station.

———. 1949. "Control of Noxious Range Plants in a Range Management Program." *Journal of Range Management* 2:128–32.

———. 1954. "Application of Ecology in the Determination of Range Condition and Trend." *Journal of Range Management* 7:14–23.

Parker, Kenneth W., W. R. Chapline, Lloyd W. Swift, George W. Craddock, Hudson G. Reynolds, Donald R. Cornelius, and Harold H. Biswell. 1967. "Arthur W. Sampson: Pioneer Range Scientist." *Journal of Range Management* 20:346–52.

Parker, Kenneth W., and Robert W. Harris. 1959. "The 3-Step Method for Measuring Condition and Trend of Forest Ranges: A Resume of Its History, Development, and Use." In *Techniques and Methods of Measuring Understory Vegetation: Proceedings of the Symposium*, 55–69. Washington, DC: USDA Forest Service.

Parker, Kenneth W., and S. Clark Martin. 1952. *The Mesquite Problem on Southern Arizona*

Ranges. Circular no. 908. Washington, DC: USDA Southwestern Forest and Range Experiment Station.

Parker, K. W., and W. G. McGinnies. 1940. *Reseeding Southwestern Ranges*. Research Note no. 86. Tucson, AZ: USDA Southwestern Forest and Range Experiment Station.

Paulsen, Harold A., and Fred N. Ares. 1961. "Trends in Carrying Capacity and Vegetation on an Arid Southwestern Range." *Journal of Range Management* 14:78–83.

Pearce, Matthew Allen. 2014. "Discontent on the Range: Uncovering the Origins of Public Grazing Lands Politics in the Intermountain West." PhD dissertation, University of Oklahoma.

Pearse, C. Kenneth. 1943. "Regrassing the Range." In *Yearbook of Agriculture*, 897–904. Washington, DC: USDA.

Pechanec, Joseph F., and George Stewart. 1944. *Sagebrush Burning: Good and Bad*. Farmers' Bulletin no. 1948. Washington, DC: USDA.

Peffer, E. Louise. 1951. *The Closing of the Public Domain: Disposal and Reservation Policies, 1900–50*. Stanford, CA: Stanford University Press.

Pellant, Mike. 1996. *Cheatgrass: The Invader That Won the West*. Washington, DC: US Department of the Interior, Bureau of Land Management.

Pellant, Michael L., Patrick Shaver, David A. Pyke, and Jeffrey E. Herrick. 2005. *Interpreting Indicators of Rangeland Health*. Version 4. Technical Reference 1734-6. Denver, CO: US Department of the Interior, Bureau of Land Management, National Science and Technology Center, Division of Science Integration, and Branch of Publishing Services.

Penny, J. Russell, and Marion Clawson. 1953. "Administration of Grazing Districts." *Land Economics* 29:23–34.

Perevolotsky, Avi, and No'am G. Seligman. 1998. "Role of Grazing in Mediterranean Rangeland Ecosystems." *Bioscience* 48:1007–17.

Peterson, David L., and V. Thomas Parker, eds. 1998. *Ecological Scale: Theory and Applications*. New York: Columbia University Press.

Pickford, G. D. 1940. "Range Survey Methods in Western United States." *Herbage Reviews* 8:1–12.

Poulton, Charles E. 1984. "Management of Pastoral Development in the Third World by Stephen Sandford." *Journal of Range Management* 37:478.

Pound, Roscoe, and Frederic E. Clements. 1898. "The Phytogeography of Nebraska: I. General Survey." Lincoln, NE: J. North.

Prevedel, David A., and Curtis M. Johnson. 2005. *Beginnings of Range Management: Albert F. Potter, First Chief of Grazing, US Forest Service, and Photographic Comparison of His 1902 Forest Reserve Survey in Utah with Conditions 100 Years Later*. Intermountain Region Report R4-VM 2005-01. Ogden, UT: USDA Forest Service.

Prevedel, David A., E. Durant McArthur, and Curtis M. Johnson. 2005. *Beginnings of Range Management: An Anthology of the Sampson-Ellison Photo Plots (1913 to 2003) and a Short History of the Great Basin Experiment Station*. General Technical Report RMRS-GTR-154. Ogden, UT: USDA Forest Service, Rocky Mountain Research Station.

Provenza, Fred D., Juan J. Villalba, L. E. Dziba, Shelton B. Atwood, and Roger E. Banner. 2003. "Linking Herbivore Experience, Varied Diets, and Plant Biochemical Diversity." *Small Ruminant Research* 49:257–74.

Pyne, Stephen J. 1982. *Fire in America: A Cultural History of Wildland and Rural Fire*. Princeton, NJ: Princeton University Press.

———. 1997. *World Fire: The Culture of Fire on Earth*. Seattle: University of Washington Press.

Pyne, Stephen J., Patricia L. Andrews, and Richard D. Laven. 1996. *Introduction to Wildland Fire*. 2nd ed. New York: Wiley.

Rakestraw, Lawrence. 1958. "Sheep Grazing in the Cascade Range: John Minto vs. John Muir." *Pacific Historical Review* 27:371–82.

Rand, Frederick V. 1945. "Agricultural Researcher and Statesman." *USDA Employee News Bulletin* 4 (6) (March 19), n.p.

Reid, Elbert H., and G. D. Pickford. 1944. "An Appraisal of Range Survey Methods." *Journal of Forestry* 42:471–79.

Reid, Robin S., Kathleen A. Galvin, and Russell S. Kruska. 2008. "Global Significance of Extensive Grazing Lands and Pastoral Societies: An Introduction." In *Fragmentation in Semi-Arid and Arid Landscapes: Consequences for Human and Natural Systems*, edited by Kathleen A. Galvin, Robin S. Reid, R. H. Behnke, and N. Thompson Hobbs, 1–24. Dordrecht: Springer.

Reisner, Marc. 1986. *Cadillac Desert: The American West and Its Disappearing Water*. New York: Viking.

Remley, David. 1993. *Bell Ranch: Cattle Ranching in the Southwest, 1824–1947*. Albuquerque: University of New Mexico Press.

Renner, Frederic G. 1948. "Range Condition: A New Approach to the Management of Natural Grazing Lands." In *Proceedings of the Inter-American Conference on Conservation of Renewable Natural Resources: Denver, Colorado, September 7–20, 1948*, 527–35. Publication 3382. Washington, DC: Department of State.

———. 1951. "Recent Advances in Methods for Restoring Deteriorated Grazing Land." In *Proceedings*, Vol. 6. United Nations Scientific Conference on the Conservation and Utilization of Resources, 544–48. Lake Success, NY: United Nations Department of Economic Affairs.

Reynolds, H. G., and J. W. Bohning. 1956. "Effects of Burning on a Desert Grass-Shrub Range in Southern Arizona." *Ecology* 37:769–77.

Reynolds, H. G., and J. G. Bridges. 1950. "Working Plan for Reseeding Research in the Southwest." Typescript located in the Records of the Forest Service (Record Group 95), National Archives and Records Administration, College Park, MD.

Reynolds, H. G., and G. E. Glendening. 1949. "Merriam Kangaroo Rat a Factor in Mesquite Propagation on Southern Arizona Range Lands." *Journal of Range Management* 2:193–97.

Reynolds, H. G., F. Lavin, and H. W. Springfield. 1949. *Preliminary Guide for Range Reseeding in Arizona and New Mexico*. Research Report 7. Tucson, AZ: USDA Southwestern Forest and Range Experiment Station.

Rice, Barbara, and Mark Westoby. 1978. "Vegetative Responses of Some Great Basin Shrub Communities Protected against Jackrabbits or Domestic Stock." *Journal of Range Management* 31:28–34.

Richards, William A., Frederick Haynes Newell, Gifford Pinchot, Albert Franklin Potter, Frederick V. Coville, and John H. Hatton, eds. 1905. *Report of the Public Lands Commission: With Appendix*. 58th Congress, 3d Session. Senate Document no. 189. Washington, DC: Government Printing Office.

Riddell, James C. 1982. *Land Tenure Issues in West African Livestock and Range Development Projects*. Research Paper 77. Madison, WI: Land Tenure Center.

Rissman, Adena R., and Nathan F. Sayre. 2012. "Conservation Outcomes and Social Relations: A Comparative Study of Private Ranchland Conservation Easements." *Society and Natural Resources* 25:523–38.

Robertson, Thomas. 2012. *The Malthusian Moment: Global Population Growth and the Birth of American Environmentalism*. New Brunswick, NJ: Rutgers University Press.

Robin, Libby. 1997. "Ecology: A Science of Empire?" In *Ecology and Empire: Environmental History of Settler Societies*, edited by T. Griffiths and L. Robin, 63–75. Seattle: University of Washington Press.

———. 2007. *How a Continent Created a Nation*. Sydney: University of New South Wales Press.

Romero, Adam. 2015. "Commercializing Chemical Warfare: Citrus, Cyanide, and an Endless War." *Agriculture and Human Values*. doi:10.1007/s10460-015-9591-1.

Roth, Filibert. 1902. "Grazing in the Forest Reserves." In *Yearbook of the Department of Agriculture 1901*, 333–48. Washington, DC: USDA.

Roundy, Bruce A., and Sharon H. Biedenbender. 1995. "Revegetation in the Desert Grassland." In *The Desert Grassland*, edited by Mitchel P. McClaran and Thomas R. Van Devender, 265–303. Tucson: University of Arizona Press.

Rowley, William D. 1985. *U.S. Forest Service Grazing and Rangelands: A History*. College Station: Texas A&M University Press.

———. 1999. "Historical Considerations in the Development of Range Science." In *Forest and Wildlife Science in America: A History*, edited by Harold K. Steen, 230–60. Durham, NC: Forest History Society.

Ruddiman, W. F. 2005. *Plows, Plagues, and Petroleum: How Humans Took Control of Climate*. Princeton, NJ: Princeton University Press.

Sampson, Arthur W. 1908. *Revegetation of Overgrazed Range Areas*. Circular no. 158. Washington, DC: USDA Forest Service.

———. 1909. *Natural Revegetation of Depleted Mountain Grazing Lands*. Circular no. 169. Washington, DC: USDA Forest Service.

———. 1914. "Natural Revegetation of Range Lands Based Upon Growth Requirements and Life History of the Vegetation." *Journal of Agricultural Research* 3:93–147.

———. 1919. *Plant Succession in Relation to Range Management*. USDA Bulletin 791. Washington, DC: Government Printing Office.

———. 1923. *Range and Pasture Management*. New York: Wiley.

———. 1951. "Application of Ecological Principles in Determining Condition of Grazing Lands." In *Proceedings*, Vol. 6. United Nations Scientific Conference on the Conservation and Utilization of Resources, 509–14. Lake Success, NY: United Nations Department of Economic Affairs.

———. 1952. *Range Management: Principles and Practices*. New York: Wiley.

Sampson, Arthur W., and Arnold M. Schultz. 1957. *Control of Brush and Undesirable Trees*. Rome: Food and Agriculture Organization of the United Nations.

Sampson, Arthur William, and Leon Henry Weyl. 1918. *Range Preservation and Its Relation to Erosion Control on Western Grazing Lands*. Bulletin no. 675. Washington, DC: USDA.

Samson, Fred B., Fritz L. Knopf, and Wayne R. Ostlie. 1998. "Grasslands." In *Status and Trends of the Nation's Biological Resources*, edited by Michael J. Mac, Paul A. Opler, Catherine E. Puckett Haecker, and Peter D. Doran, Vol. 2, 437–72. Reston, VA: US Department of Interior, US Geological Survey.

Sandford, Stephen. 1976. "Pastoralism under Pressure." *Development Policy Review* 9:45–68.

———. 1981. "Review of World Bank Livestock Activities in Dry Tropical Africa." Unpublished consultant's report. Washington, DC: World Bank.

———. 1982. "Pastoral Strategies and Desertification: Opportunism and Conservatism in Dry Lands." In *Desertification and Development: Dryland Ecology in Social Perspective*, edited by B. Spooner and H. S. Mann, 61–80. London: Academic Press.

———. 1983. *Management of Pastoral Development in the Third World.* Chichester, UK: Wiley.

Sauer, Carl O. 1950. "Grassland Climax, Fire, and Man." *Journal of Range Management* 3:16–21.

Savory, Allan. 1978. "A Holistic Approach to Ranch Management Using Short Duration Grazing." In *Proceedings of the First International Rangeland Congress. Denver, Colorado,* edited by Donald N. Hyder, 555–57. Denver, CO: Society for Range Management.

———. 1983. "The Savory Grazing Method or Holistic Resource Management." *Rangelands* 5:155–59.

Savory, Allan, and Stanley D. Parsons. 1980. "The Savory Grazing Method." *Rangelands* 2:234–37.

Sayer, R. Andrew. 1992. *Method in Social Science: A Realist Approach.* 2nd ed. London: Routledge.

Sayre, Nathan F. 1999. "The Cattle Boom in Southern Arizona: Towards a Critical Political Ecology." *Journal of the Southwest* 41:239–71.

———. 2001. *The New Ranch Handbook: A Guide to Restoring Western Rangelands.* Santa Fe, NM: Quivira Coalition.

———. 2002. *Ranching, Endangered Species, and Urbanization in the Southwest: Species of Capital.* Tucson: University of Arizona Press.

———. 2003. "Recognizing History in Range Ecology: 100 Years of Science and Management on the Santa Rita Experimental Range." In *Santa Rita Experimental Range: 100 Years (1903–2003) of Accomplishments and Contributions,* edited by Mitchel P. McClaran, Peter F. Ffolliott, and Carleton B. Edminster, 1–15. Proc. RMRS-P-30. Ogden, UT: USDA Rocky Mountain Research Station.

———. 2005a. "Ecological and Geographical Scale: Parallels and Potential for Integration." *Progress in Human Geography* 29:276–90.

———. 2005b. "Interacting Effects of Landownership, Land Use, and Endangered Species on Conservation of Southwestern US Rangelands." *Conservation Biology* 19:783–92.

———. 2005c. *Working Wilderness: The Malpai Borderlands Group and the Future of the Western Range.* Tucson, AZ: Rio Nuevo Press.

———. 2008. "The Genesis, History, and Limits of Carrying Capacity." *Annals of the Association of American Geographers* 98:120–34.

———. 2009. "Scale." In *A Companion to Environmental Geography,* edited by Noel Castree, David Demeritt, Bruce Rhoads, and Diana Liverman, 95–108. Chichester, UK: Wiley-Blackwell.

———. 2010. "Climax and 'Original Capacity': The Science and Aesthetics of Ecological Restoration in the Southwestern USA." *Ecological Restoration* 28:23–31.

———. 2012. "The Politics of the Anthropogenic." *Annual Review of Anthropology* 41:57–70.

———. 2015. "Scales and Polities." In *Routledge Handbook of Political Ecology,* edited by Tom Perreault, Gavin Bridge, and James McCarthy, 504–15. London: Routledge.

Sayre, Nathan F., Eric Biber, and Greta Marchesi. 2013. "Social and Legal Effects on Monitoring and Adaptive Management: A Case Study of National Forest Grazing Allotments, 1927–2007." *Society and Natural Resources* 26:86–94.

Sayre, Nathan F., William deBuys, Brandon Bestelmeyer, and Kris Havstad. 2012. "'The Range Problem' after a Century of Rangeland Science: New Research Themes for Altered Landscapes." *Rangeland Ecology and Management* 65:545–52.

Sayre, Nathan F., and Alan Di Vittorio. 2009. "Scale." In *The International Encyclopedia of Human Geography,* edited by Robert Kitchin and Nigel Thrift, vol. 1, 19–28. Oxford: Elsevier.

Sayre, Nathan F., Ryan R.J. McAllister, Brandon T. Bestelmeyer, Mark Moritz, and Matthew D. Turner. 2013. "Earth Stewardship of Rangelands: Coping with Ecological, Economic, and Political Marginality." *Frontiers in Ecology and the Environment* 11: 348–54.

Scarnecchia, David L. 1999. "The Range Utilization Concept, Allocation Arrays, and Range Management Science." *Journal of Range Management* 52:157–60.

Scott, J. Michael, Frank W. Davis, R. Gavin McGhie, R. Gerald Wright, Craig Groves, and John Estes. 2001. "Nature Reserves: Do They Capture the Full Range of America's Biological Diversity?" *Ecological Applications* 11:999–1007.

Seager, Richard, Mingfang Ting, Isaac Held, Yochanan Kushnir, Jian Lu, Gabriel Vecchi, Huei-Ping Huang, et al. 2007. "Model Projections of an Imminent Transition to a More Arid Climate in Southwestern North America." *Science* 316:1181–84.

Seager, Richard, and Gabriel A. Vecchi. 2010. "Greenhouse Warming and the 21st Century Hydroclimate of Southwestern North America." *Proceedings of the National Academy of Sciences* 107:21277–82.

Sears, Paul B. 1935. *Deserts on the March*. Norman: University of Oklahoma Press.

Secretary of Agriculture, ed. 1933. *A National Plan for American Forestry*. 73d Cong., 1st Sess. Senate. Doc. 12. Washington, DC: Government Printing Office.

———. 1936. *The Western Range. Letter from the Secretary of Agriculture Transmitting in Response to Senate Resolution No. 289: A Report on the Western Range—A Great but Neglected Natural Resource.* Washington, DC: Government Printing Office.

Seligman, No'am G., Eddy van der Maarel, and Sandra Díaz. 2011. "Imanuel Noy-Meir—The Ecologist and the Man." *Israel Journal of Ecology and Evolution* 57:5–16.

Shapiro, Aaron. 2014. "A Grand Experiment: USDA Forest Service Experimental Forests and Ranges." In *USDA Forest Service Experimental Forests and Ranges*, edited by Deborah C. Hayes, Susan L. Stout, Ralph H. Crawford, and Anne P. Hoover, 3–23. New York: Springer.

Shear, Cornelius Lott. 1901. *Field Work of the Division of Agrostology: A Review and Summary of the Work Done since the Organization of the Division, July 1, 1895*. Bulletin 25. Washington, DC: USDA Division of Agrostology.

Sheffer, Theo. H. 1911. "The Prairie-Dog Situation in Kansas." *Transactions of the Kansas Academy of Sciences* 23/24:115–18.

Sierra-Corona, Rodrigo, Ana Davidson, Ed L. Fredrickson, Hugo Luna-Soria, Humberto Suzan-Azpiri, Eduardo Ponce-Guevara, and Gerardo Ceballos. 2015. "Black-Tailed Prairie Dogs, Cattle, and the Conservation of North America's Arid Grasslands." *PLoS ONE* 10: e0118602.

Sills, Peter. 2014. *Toxic War: The Story of Agent Orange*. Nashville, TN: Vanderbilt University Press.

Slotkin, Richard. 1985. *The Fatal Environment: The Myth of the Frontier in the Age of Industrialization, 1800–1890*. New York: Atheneum.

Smith, Frederick E. 1968. "The International Biological Program and the Science of Ecology." *Proceedings of the National Academy of Sciences of the United States of America* 60:5–11.

Smith, Jared G. 1895. *A Note on Experimental Grass Gardens*. USDA Division of Agrostology Circular no. 1. Washington, DC: Government Printing Office.

———. 1898. *Experiments in Range Improvements*. USDA Division of Agrostology Circular no. 8. Washington, DC: Government Printing Office.

———. 1899. *Grazing Problems in the Southwest and How to Meet Them*. USDA Division of Agrostology Bulletin no. 16. Washington, DC: Government Printing Office.

Smith, Lamar, George Ruyle, Jim Maynard, Steve Barker, Walt Meyer, Dave Stewart, Bill Coulloudon, Stephen Williams, and Judith Dyess. 2005. *Principles of Obtaining and Interpreting Utilization Data on Southwest Rangelands*. Report AZ1375. Tucson: University of Arizona Cooperative Extension.

Smith, Neil. 1984. *Uneven Development: Nature, Capital, and the Production of Space*. Cambridge: Basil Blackwell.

Soulé, Michael E. 1985. "What Is Conservation Biology? A New Synthetic Discipline Addresses the Dynamics and Problems of Perturbed Species, Communities, and Ecosystems." *BioScience* 35:727–34.

Starrs, Paul F. 1998. *Let the Cowboy Ride: Cattle Ranching in the American West*. Baltimore: Johns Hopkins University Press.

Steen, Harold K. 1976. *The U.S. Forest Service: A History*. Seattle: University of Washington Press.

———. 1998. *Forest Service Research: Finding Answers to Conservation's Questions*. Durham, NC: Forest History Society.

Stegner, Wallace. 1954. *Beyond the Hundredth Meridian: John Wesley Powell and the Second Opening of the West*. Boston: Houghton Mifflin.

Stoddart, L. A. 1935. "Range Capacity Determination." *Ecology* 16:531–33.

———. 1953. "Problems in Estimating Grazing Capacity of Ranges." In *Proceedings of the 6th International Grassland Congress*, 1367–73. Washington, DC: National Publishing.

Stoddart, L. A., and Arthur D. Smith. 1943. *Range Management*. 1st ed. New York, London: McGraw-Hill.

Stoddart, L. A., Arthur D. Smith, and Thadis W. Box. 1975. *Range Management*. 3d ed. New York: McGraw-Hill.

Streets, Rubert B., and Ernest B. Stanley. 1938. *Control of Mesquite and Noxious Shrubs on Southern Arizona Grassland Ranges*. Technical Bulletin no. 74. Tucson: Arizona Agricultural Experiment Station.

Stromberg, Mark R., Jeffrey D. Corbin, and Carla Marie D'Antonio. 2007. *California Grasslands: Ecology and Management*. Berkeley: University of California Press.

Sullins, Martha J., David T. Theobald, Jeff R. Jones, and Leah M. Burgess. 2002. "Lay of the Land: Ranch Land and Ranching. In *Ranching West of the 100th Meridian: Culture, Ecology, and Economics*, edited by Richard L. Knight, Wendell C. Gilgert, and Ed Marston, 25–31. Washington, DC: Island Press.

Svejcar, Tony, and Joel R. Brown. 1991. "Failures in the Assumptions of the Condition and Trend Concept for Management of Natural Ecosystems." *Rangelands* 13:165–67.

Swift, Jeremy. 1977a. "Desertification and Man in the Sahel." In *Landuse and Development*, edited by Phil O'Keefe and Ben Wisner, 171–78. London: International African Institute.

———. 1977b. "Sahelian Pastoralists: Underdevelopment, Desertification, and Famine." *Annual Review of Anthropology*, 457–78.

———. 1994. "Dynamic Ecological Systems and the Administration of Pastoral Development." In *Living with Uncertainty: New Directions in Pastoral Development in Africa*, edited by Ian Scoones, 153–73. London: Intermediate Technology Publications.

Swyngedouw, Erik. 1997. "Neither Global nor Local: 'Glocalization' and the Politics of Scale." In *Spaces of Globalization: Reasserting the Power of the Local*, edited by Kevin R. Cox, 137–66. New York: Guilford Press.

Talbot, M. W. 1937. *Indicators of Southwestern Range Conditions*. Farmers' Bulletin no. 1782. Washington, DC: USDA.

Talbot, M. W., and F. F. Cronemiller. 1961. "Some of the Beginnings of Range Management." *Journal of Range Management* 14:95–102.

Tansley, A. G. 1923. *Practical Plant Ecology: A Guide for Beginners in Field Study of Plant Communities*. London: Allen & Unwin.

Taylor, Peter J. 1992. "Re/constructing Socioecologies: System Dynamics Modeling of Nomadic Pastoralists in Sub-Saharan Africa." In *The Right Tools for the Job: At Work in Twentieth-Century Life Sciences*, edited by Adele E. Clark and Joan H. Fujimura, 115–48. Princeton, NJ: Princeton University Press.

Taylor, Walter P. 1930. "Methods of Determining Rodent Pressure on the Range." *Ecology* 11:523–42.

Taylor, Walter P., and J. V. G. Loftfield. 1924. *Damage to Range Grasses by the Zuni Prairie Dog*. USDA Department Bulletin no. 1227. Washington, DC: Government Printing Office.

Thomas, David S. G., and Nick Middleton. 1994. *Desertification: Exploding the Myth*. Chichester, UK: Wiley.

Tobey, Ronald C. 1981. *Saving the Prairies: The Life Cycle of the Founding School of American Plant Ecology, 1895–1955*. Berkeley: University of California Press.

Tootel, R. B. 1936. *Grazing Districts: Their Nature and Possibilities in Range Land Utilization*. Bulletin no. 127. Montana Extension Service in Agriculture and Home Economics.

Torell, L. A., N. R. Rimbey, O. A. Ramirez, and D. W. McCollum. 2005. "Income Earning Potential versus Consumptive Amenities in Determining Ranchland Values." *Journal of Agricultural and Resource Economics* 30:537–60.

Tschirley, Fred H. 1968. *An Assessment of Ecological Consequences of the Defoliation Program in Vietnam*. Saigon.

Tschirley, Fred H., and Herbert M. Hull. 1959. "Susceptibility of Velvet Mesquite to an Amine and an Ester of 2, 4, 5-T as Related to Various Biological and Meteorological Factors." *Weeds* 7:427–35.

United Nations Scientific Conference on the Conservation and Utilization of Resources (UNSCCUR). 1950. *Proceedings*, vol. 1. Lake Success, NY: United Nations Department of Economic Affairs.

United States Agency for International Development (USAID). 1972. *Desert Encroachment on Arable Lands: Significance, Causes and Control*. Washington, DC: Office of Science and Technology, Agency for International Development.

Upson, Arthur, W. J. Cribbs, and E. B. Stanley. 1937. *Occurrence of Shrubs on Range Areas in Southeastern Arizona*. Tucson: Arizona Agricultural Experiment Station.

USDA. 1936. *Atlas of American Agriculture: Physical Basis, Including Land Relief, Climate, Soils, and Natural Vegetation*. Washington, DC: Government Printing Office.

———. 1952. *The Santa Rita Experimental Range*. Tucson, AZ: USDA Forest Service, Southwestern Forest and Range Experiment Station.

———. 2010. *Sheep and Goats Death Loss*. Washington, DC: USDA National Agricultural Statistics Service.

———. 2011. *Cattle Death Loss*. Washington, DC: USDA National Agricultural Statistics Service.

US Forest Service. 1922. "District 3 Grazing Studies Conference." Mimeographed report located in the archives of the Santa Rita Experimental Range, Tucson, AZ.

———. 1923. "Minutes of the Grazing Conference held at Ogden, March 4–12, 1923." Records of the Forest Service (Record Group 95), National Archives and Records Administration, College Park, MD.

———. 1939. *Proceedings of Range Research Seminar*. Ogden, UT: USDA Forest Service.

Vallentine, John F. 1979. "The Literature of Range Science Based on Citations in the Journal of Range Management." *Journal of Range Management* 32:241–43.

Van Voorthuizen, E. G. 1978. "Global Desertification and Range Management: An Appraisal." *Journal of Range Management* 31:378–80.

Vasey, George. 1886. *Report of an Investigation of the Grasses of the Arid Districts of Kansas, Nebraska, and Colorado*. Bulletin 1. Washington, DC: USDA Botanical Division.

Vesk, Peter A., and Mark Westoby. 2001. "Predicting Plant Species' Responses to Grazing." *Journal of Applied Ecology* 38:897–909.

Vetter, Susanne. 2005. "Rangelands at Equilibrium and Non-Equilibrium: Recent Developments in the Debate." *Journal of Arid Environments* 62:321–41.

Vogt, William. 1948. *Road to Survival*. New York: W. Sloane.

Wagner, Joe A. 1949. "Results of Airplane Pellet Seeding on Indian Reservations." *Journal of Forestry* 47:632–35.

Wagner, R. E., W. M. Myers, S. H. Gaines, S. S. Atwood, R. L. Lovvorn, W. R. Chapline, F. G. Renner, W. A. Dayton, F. E. Bear, C. S. Garrison, G. F. Brown, R. E. Hodgson, N. R. Ellis, F. W. Duffee, H. L. Lucas, and S. H. Work, eds. 1952. "Foreword." In *Proceedings of the Sixth International Grassland Congress*, Vol. 1, v–vi. State College: Pennsylvania State College.

Walker, Brian H., Donald Ludwig, Crawford S. Holling, and Richard M. Peterman. 1981. "Stability of Semi-Arid Savanna Grazing Systems." *Journal of Ecology* 69:473–98.

Walker, Brian, and Mark Westoby. 2011. "States and Transitions: The Trajectory of an Idea, 1970–2010." *Israel Journal of Ecology and Evolution* 57:17–22.

Walters, Carl J., and C. S. Holling. 1990. "Large-Scale Management Experiments and Learning by Doing." *Ecology* 71:2060–68.

Webb, Walter Prescott. 1931. *The Great Plains*. Boston: Ginn.

Weisiger, Marsha L. 2009. *Dreaming of Sheep in Navajo Country*. Seattle: University of Washington Press.

Welch, B. L. 2005. *Big Sagebrush: A Sea Fragmented into Lakes, Ponds, and Puddles*. General Technical Report RMRS-GTR-144. Fort Collins, CO: USDA Forest Service, Rocky Mountain Research Station.

Weltzin, Jake F., Steve Archer, and Rod K. Heitschmidt. 1997. "Small-Mammal Regulation of Vegetation Structure in a Temperate Savanna." *Ecology* 78:751–63.

West, Neil E. 2003. "History of Rangeland Monitoring in the U.S.A." *Arid Land Research and Management* 17:495–545.

Westoby, Mark. 1979. "Elements of a Theory of Vegetation Dynamics in Arid Rangelands." *Israel Journal of Botany* 28:169–94.

Westoby, Mark, Brian Walker, and Imanuel Noy-Meir. 1989. "Opportunistic Management for Rangelands Not at Equilibrium." *Journal of Range Management* 42:266–74.

White, Richard. 1983. *The Roots of Dependency: Subsistence, Environment, and Social Change among the Choctaws, Pawnees, and Navajos*. Lincoln: University of Nebraska Press.

———. 1991. *"It's Your Misfortune and None of My Own": A History of the American West*. Norman: University of Oklahoma Press.

Widstrand, Carl Gösta. 1975. "The Rationale of Nomad Economy." *Ambio* 4:146–53.

Wiens, John A. 1977. "On Competition and Variable Environments." *American Scientist* 65:590–97.

———. 1984. "On Understanding a Non-Equilibrium World: Myth and Reality in Community Patterns and Processes." In *Ecological Communities: Conceptual Issues and the Evidence*, edited by Donald R. Strong, Daniel Simberloff, Lawrence G. Abele, and Ann B. Thistle, 439–57. Princeton, NJ: Princeton University Press.

Williamson, James Alexander, Clarence King, Alexander Thompson Britton, Thomas Donaldson, and John Wesley Powell, eds. 1880. *Report of the Public Lands Commission*. 46th Cong., 2d Sess. House. Ex. Doc. 46. Washington DC: Government Printing Office.

Williams, Thomas A. 1896. *Renewing of Worn-Out Native Prairie Pastures*. USDA Division of Agrostology Circular no. 4. Washington, DC: Government Printing Office.

Wilson, C. P. 1931. *Artificial Reseeding of New Mexico Ranges*. Bulletin no. 189. State College: New Mexico College of Agriculture and Mechanic Arts.

Wooton, E. O. 1908. *The Range Problem in New Mexico*. New Mexico College of Agriculture and Mechanic Arts Agricultural Experiment Station Bulletin no. 66. Albuquerque: Albuquerque Morning Journal.

Worster, Donald. 1979. *Dust Bowl: The Southern Plains in the 1930s*. New York: Oxford University Press.

———. 1985. *Rivers of Empire: Water, Aridity, and the Growth of the American West*. New York: Pantheon Books.

———. 1994. *Nature's Economy: A History of Ecological Ideas*. 2nd ed. Cambridge: Cambridge University Press.

Wu, Jianguo, and O. L. Loucks. 1995. "From Balance of Nature to Hierarchical Patch Dynamics." *Quarterly Review of Biology* 70:439–66.

Wuerthner, George, and Mollie Matteson, eds. 2002. *Welfare Ranching: The Subsidized Destruction of the American West*. Washington, DC: Island Press.

Young, James A. 2000. "Range Research in the Far Western United States: The First Generation." *Journal of Range Management* 53:2–11.

Young, James A., and Fay L. Allen. 1997. "Cheatgrass and Range Science: 1930–1950." *Journal of Range Management* 50:530–35.

Young, James A., and Charlie D. Clements. 2001. Range Research: The Second Generation. *Journal of Range Management* 54:115–21.

———. 2009. *Cheatgrass: Fire and Forage on the Range*. Reno: University of Nevada Press.

Young, Stanley P. 1936. *Rodent Control Aided by Emergency Conservation Work*. Wildlife Research and Management Leaflet BS-54. Washington, DC: USDA Bureau of Biological Survey.

INDEX

Page numbers in italics refer to figures and tables.